C for Engineers

Kenneth W. Collier
Department of Computer Science and Engineering
Northern Arizona University
Flagstaff, Arizona

The Benjamin/Cummings Publishing Company, Inc.

Redwood City, California · Menlo Park, California
Reading, Massachusetts · New York · Don Mills, Ontario
Wokingham, U.K. · Amsterdam · Bonn · Singapore
Tokyo · Madrid · San Juan

The author and publisher gratefully acknowledge the
contributions of the following individuals who
reviewed the manuscript: Stormy Attaway, Boston
University; Scott Iverson, University of Washington;
John E. Parsons, North Carolina State University;
and Ronald E. Lacey, Texas A & M University. We
especially thank Dr. Lacey for his contributions to the
problem sets throughout the text.

The examples presented in this book have been
included for their instructional value. They have been
tested with care but are not guaranteed for any
particular purpose. The publisher does not offer
warranties or representations, nor does it accept any
liabilities with respect to the programs.

This is a module in *The Engineer's Toolkit*, a
Benjamin/Cummings SELECT edition. Contact your
sales representative for more information.

The *Engineer's Toolkit* and SELECT are trademarks of
the Benjamin/Cummings Publishing Company, Inc.
Contact your sales representative for more
information.

Photo Credits:
Chapter 1: NASA-LBJ Space Center
Chapter 2: Courtesy of Lockheed
Chapter 3: NASA-LBJ Space Center
Chapter 4: © Mehau Kulyk/Science Photo Library/
          Photo Researchers, Inc.
Chapter 5: Courtesy of Boeing Corporation
Chapter 6: Courtesy of General Motors
Chapter 7: Courtesy of Wolfram Research, Inc.

ISBN 0-8053-6476-5

The Benjamin/Cummings Publishing Company, Inc.
390 Bridge Parkway
Redwood City, CA 94065

Contents

1 An Introduction to Problem Solving in C

In 1977, NASA sent two Mariner spacecraft named Voyager 1 and Voyager 2 to Jupiter and Saturn. The two spacecraft encountered Jupiter in 1979 and Saturn in 1980 and 1981. Voyager 1 then began moving out of the plane of the solar system. Voyager 2 passed and photographed Uranus and Neptune before heading out of the solar system's plane. In order for the Voyager 2 to travel as far as it did, engineers had to solve a myriad of problems, such as keeping the probe within the solar system, providing robust and reliable communications between the probe and Earth, and designing the probe to withstand the unknown space environments that the craft would encounter. The Voyager project is heralded as a highly successful engineering effort that required the problem-solving expertise and collaborative efforts of talented professionals from many different engineering disciplines.

INTRODUCTION

This chapter will introduce a problem-solving methodology that can be used to solve computer-programming problems regardless of the programming language used. This methodology will be modeled using the *C* programming language, but it is important to understand that the methodology is not limited to programming in *C*. For that matter, this problem-solving process is not limited to programming at all. Programming is used to solve a class of problems that fit within a larger class known as open-ended problems. *Open-ended problems* have a variety of valid solutions. These problems are interesting because they stimulate the creative talents of the problem solver. However, open-ended problems are often the most difficult to solve. Although most engineering problems are open ended, the same basic problem-solving methodology can be applied to both open-ended and closed-ended engineering problems.

Throughout this text, short exercises called Try It! exercises will allow you to check quickly to see if you have understood the materials just presented. Additionally, What If? sections will encourage you to extend the ideas and examples presented in the text. You will understand the material in this text much better if you take a little extra time to solve these problems.

1-1   WHAT IS ANSI *C*?

The first digital computers (ABC, ENIAC, UNIVAC, and so on) were programmed using switches or by reconfiguring hardware connections. Programs were written in *machine language,* the computer's native language that represents commands and data as binary (base 2) values. Programming was very tedious and time consuming, and logical errors were very costly to fix. In 1957, John Backus completed the development of the first high-level language known as FORTRAN I. A *high-level language* is a programming language that allows the programmer to write programs using an English-like grammar rather than binary codes. A compiler is then used to translate these programs into machine language. Since 1957, many high-level languages have been developed, each intended to serve a certain purpose.

C is a high-level programming language that was developed in the early 1970s at Bell Laboratories by Dennis Ritchie. Its roots lie in other languages such as ALGOL 60 (1960), CPL (1963), BCPL (1967), and a language developed by Ken Thompson at Bell Labs in 1970 called B. Originally, *C* was used for the development of operating systems and general systems programming. Because of its use in these applications, *C* is sometimes referred to as a mid-level language.

Over the past two decades *C* has grown dramatically in popularity for general-purpose commercial programming. This popularity is due to several factors. *C* allows the programmer a lot of flexibility in manipulating various parts of the computer and the operating system. *C* is also a relatively small language in terms of its built-in syntactical elements. *Syntactical elements* include the keywords, operators, and punctuation marks that form the building blocks of a language. However, *C* allows the creation and use of *libraries* of "tools" that can be used in several different programs. This notion of libraries allows the language to expand to suit the needs of most programmers. *C* programs are very efficient and have fast execution times. They are also very portable. This means that a *C* program that is developed on one computer is likely to run correctly on other computers.

Originally, the *C* language was not standardized; that is, there was not a precise, detailed, and widely accepted description of the language. The first move toward a standard was published in 1978 in *The C Programming Language* by Brian Kernighan and Dennis Ritchie. It was quickly accepted as the de facto standard for the *C* language and became known as K&R *C*. One problem with de facto language standards is that there is no one to police the continual evolution of the language. In the case of K&R *C*, many implementations had small variations and improvements making them almost, but not quite, portable.

In 1982, the *American National Standards Institute (ANSI)* formed a sub-committee charged with the official standardization of the *C* programming language. The standard, which is now referred to as *ANSI C,* was formally adopted in 1989 as American National Standard X3.159-1989. This document describes a detailed specification for the *C* language as well as a set of run-time libraries. ANSI *C* has proved to be an improvement over K&R *C* and has helped ensure portability. This text will focus on ANSI *C*. Currently, there is a growing trend toward the usage of a language called *C++* for the development of commercial software systems. *C++* is an object-oriented extension of *C* developed by Bjarne Stroustrup at AT&T's Bell Laboratories. *Object-oriented languages* provide greater programming power for solving large, complex problems than traditional programming languages.

Programming is problem solving; therefore, it is highly beneficial for you to apply an effective problem-solving methodology to programming activities. However, programming adds an extra dimension to this problem-solving process, namely, that of learning the language necessary to produce a working program that can be interpreted and executed by the computer. Learning a programming language is very much like learning a foreign language. There are rules for the correct use of syntactical elements—letters, symbols, and punctuation marks—as well as rules for the *semantics* or meaning of the commands that make up the program. Many fledgling programmers are unable to recognize that the primary challenge in programming is the problem-solving aspect and that the secondary difficulty is the mastery of the language itself. The language is simply a medium for communicating to the computer a problem solution. Learning the language elements of a programming language should not be separate from learning to solve problems, just as learning the language elements of French should not be separate from learning to communicate to French-speaking people. Although this module will introduce you to the syntactical elements of the *C* programming language, it will focus largely on the problem-solving methodology you will use to solve problems.

1-2 CREATING PROGRAM FILES AND USING COMPILERS

The computer by itself is incapable of understanding and executing a *C* program. It understands only instructions that are encoded as binary values known as machine instructions. The computer's native language is a

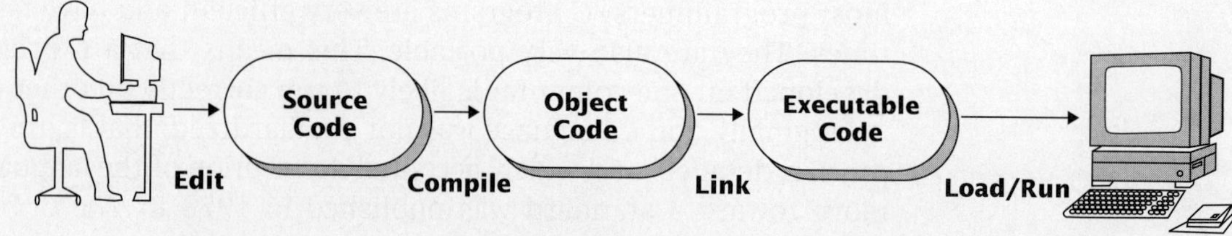

Figure 1-1 The Compilation Sequence

low-level language called machine language. A *compiler* is a computer tool whose purpose is to translate a high- or mid-level language like *C* into machine language for processing by the computer. Many *C* compilers are available for today's computers. This textbook is not specific to any particular compiler; therefore, you will need to use your compiler's manual to learn its details.

It is important to understand something about the compilation process. There are several phases in this process, which begins after the third step in the problem-solving process.

The first step in the compilation process is the creation of the source code. During this phase, the computer program is written using the proper syntax of a language such as *C*. This program is developed with the help of a text editor and is stored in an electronic file or a collection of files. Although the files may have any name the programmer chooses, it is customary to give the file a base identifier followed by a .C extension. So, a program that calculates the velocity of an unladen African swallow in flight might be stored in a source code file named SWALLOW-VELOCITY.C. Of course, you will have to adhere to the file-naming restrictions of your computer.

Once the source code is complete it is compiled into a separate file that contains relocatable object code, which is functionally equivalent to the source code but is in machine-language syntax. Files that contain relocatable object code are often denoted by a .O extension. So, our swallow program might be stored in a file called SWALLOW__VELOCITY.O after this phase.

Recall that *C* is a library-based language. It is probable that a *C* program makes use of program modules that reside in separate libraries. The next phase in the compilation process is known as linking. This phase resolves any references to library utilities. The result of this phase is a file known as a run unit or an executable file.

Next is the loading phase. This phase serves the purpose of allocating memory locations for an execution of the program.

Finally, the program is ready to be run. Many *C* compilers greatly simplify this process by offering techniques for combining the compile–link sequence and the load–run sequence. The entire sequence is graphically represented in Figure 1-1.

1-3 USING A FIVE-STEP PROBLEM-SOLVING PROCESS

All people solve a variety of problems every day. These range from such innocuous problems as how to travel from one geographical location to another to more critical problems such as how to pay the bills without causing a negative bank balance. Although engineers are interested in more analytical problems, the fact remains that they too are problem solvers. The computer's powerful problem-solving ability has made it an invaluable tool in the engineer's workspace. Unlike other electrical or mechanical devices, engineers can program computers to solve a wide variety of problems. In order to do so, computers require a very precise and detailed set of problem-solving directions.

Humans are capable of storing a vast amount of world knowledge and applying it to the solution of new problems. As an example, consider trying to explain to someone how to change a flat tire. It is probably sufficient to give such directions as, "Use the lug wrench to remove the lug nuts." The average person can probably apply his or her understanding of wrenches and nuts to perform this task. This type of instruction, however, is far too vague and imprecise for a computer to interpret and act on.

A computer stores a very small set of built-in operations that act as the building blocks for the development of the more complex operations necessary to solve interesting problems. A computer program must provide a very precise and detailed description of the steps required in a problem solution in order to be processed. However, computers are extremely fast at solving many of the problems that are very tedious and time consuming for humans to solve. So, although people must be very precise in writing programs, the computer will reward them by solving problems far more quickly than can be done by hand.

This discrepancy between human problem solvers and computers creates a difficulty for programmers. Programmers must learn how to reduce their thought processes to the very low-level details that are required to write an effective program. They must force themselves to think at a level that is much more precise and detailed than would be necessary if they were explaining the process to another human, which is exactly the reason that programs contain errors and that many people find programming difficult. An organized and well-defined problem-solving approach can go a long way toward bridging this gap. Such an approach will guide the programmer in methodically removing the layers of abstraction in understanding a given problem so that the solution can be stated with exactness and precision.

The following five-step problem-solving process will form the basis for all the topics presented in this text. It forms a model for solving a variety of problems. As you read these steps, try to see how you have applied them to past problems that you have solved. They form a natural progression in the process of solving problems.

A General Problem-Solving Procedure	A Problem-Solving Procedure for Computer Programming
1. Define the problem.	Understand what the program is supposed to do.
2. Gather information.	Determine the program's inputs and outputs.
3. Generate and evaluate potential solutions.	Design a structure and an algorithm for the program.
4. Refine and implement a solution.	Write the computer program.
5. Verify and test the solution.	Compile, test, and debug the program.

Several activities in this process must be completed before the programmer even begins writing the program. These activities are important to the success of the computer program and should not be ignored.

Try It Consider the following typical problems that might be encountered by the average person on any given day. Do each of the steps apply to the solution of these types of problems?

♦ Traveling from 123 N. Fourth St., Anywhere, United States, to 21 Downing Street, London, England.
♦ Returning defective merchandise to a retail store without having kept the receipt.

Now consider some engineering problems. Do each of the five steps apply to these types of problems?

♦ Building a device to constantly monitor the temperature of rabbits during a scientific experiment.
♦ Building a means of transporting the pieces of a Boeing 777 aircraft from Wichita, Kansas, to the assembly plant in Seattle, Washington.

Now consider some programming problems. Do each of the five steps apply to these types of problems?

♦ Developing a software system that will determine the proper missile trajectory for a given target based on wind direction, wind speed, humidity, and barometric pressure.
♦ Developing a software system that will monitor the traffic flow at an intersection and will cycle the traffic lights in order to avoid congestion of traffic in any direction.

1-4 CALCULATING THE MEAN

The five-step problem-solving process and the compilation process can be demonstrated by considering the problem of calculating the average of a collection of values. In this section we develop a working computer program to do this. Although you have not been introduced to the syntax of C, you should examine this simple program and see which parts of it are self-explanatory.

Engineers and scientists very often calculate statistics for collections of data. One simple statistic is the mean of a population, also known as the average. In this section we will work through the five-step problem-solving process toward the development of a C program that will calculate the average of a collection of real numbers.

Step 1 of the process defines the problem. The goal of this phase is to understand what the program is supposed to do, which may involve interviews with the end user of the program to establish exactly what is required. Imagine that after such an interview, the following program requirements have been established:

- The program will read a collection of positive real numbers that the user enters.
- The collection of numbers is of unknown size.
- The program will print the average of this collection on the user's display.

Step 2 consists of information gathering. The goal of this stage is to determine the program's inputs and outputs. We can infer from the problem definition that the user will enter an unlimited number of positive real numbers using the keyboard and that the user will enter a zero to terminate input. This use of zero works in this case only because the actual data includes only nonzero values. This decision could easily be replaced by another means of terminating the input.

We can also infer from the problem statement that the program will display the average of the user's collection of numbers on the computer's display screen as a real number.

Step 3 involves generating and evaluating potential solutions. The objective in this step is to design a structure and an algorithm for the program. One approach to generating a potential solution is to precisely describe how a human would solve the same problem. The outline in Example 1-1 provides a verbal description of the process of calculating averages.

EXAMPLE 1-1

An Algorithm for Calculating Averages

```
Calculate the sum of all of the numbers
Count all of the numbers
Divide the sum by the count
Print the average
```

This is known as an *algorithm*. An algorithm is a step-by-step solution to a given problem. As mentioned previously, computers must be given a very precise set of instructions at a very low level in order to work correctly. Although this algorithm makes sense to humans, it is not possible to tell the computer to "calculate the sum of all of the numbers." This must be refined, or broken into more detailed steps, in order to be useful as a computer program. The calculation of the sum of all numbers might be solved by reading in each number and adding it to a running total until the user enters a 0. The algorithm of Example 1-2 reflects this refinement.

EXAMPLE 1-2

Algorithm Refinement

```
Calculate the sum of all of the numbers
    Get a number from the user
    While the number is not a zero
        Add the number to the total
        Add one to the number count
        Get the next number
Divide the sum by the count
Print the average
```

The indentation is provided to show the refinement of a step into a set of more specific steps. The while statements demonstrate the need to repeat some steps for an indefinite number of cycles. This is known as a *while loop.* Notice that the counting of the numbers is now taken care of inside the while loop. In the first algorithm it was not clear how this step was to be accomplished. Our algorithm now reflects a much more precise set of steps for calculating the average of a set of values. However, there is still a problem with this solution. The algorithm adds a number to the total and adds 1 to the count of numbers. Yet neither of these values has a meaningful initial value. What does it mean to add 1 to something that has no value? In both of these cases, the initial value must be set to 0, and then the new values must be added in the loop. The algorithm in Example 1-3 reflects these changes.

EXAMPLE 1-3   ## Final Algorithm Refinement

```
Set total to 0
Set count to 0
Calculate the sum of all of the numbers
      Get the first number
      While number ≠ 0
            Set total to total + number
            Set count to count + 1
            Get the next number
      Endwhile
average = total/count
Print average
```

This version of our algorithm contains some other differences from the previous version. First, the use of some simple mathematical notation has made the meaning of each statement much more clear. Second, we can see the need to refer to values by a name because we do not know what the actual value is going to be. In the program these names will be used to store the values as the calculations are being performed. They are known as *variables.* These variables are represented in boldface type.

We have now generated a solution to the problem. Additionally, through the process of evaluating our solution, we have made some refinements and improvements. The problem is now ready to move into the next phase in the process.

Step 4 refines and implements the solution. Recall that the objective in this step is to write the program. The following is a short *C* program that implements the algorithm in Example 1-3. Don't worry about understanding all the details at this point. You will be learning about each of these as you read the remainder of this text.

Although the syntactical details of *C* are not yet familiar to you, if you study this example carefully, you should be able to see the similarities between this program and the algorithm developed in Example 1-3.

```c
/*------------------------------------------------------------------
   Description: This program will read positive real numbers from
       the user's keyboard, calculate the mean of the values and
       display it on the computer display.
   Programmer: Ken Collier
   Date of Last Revision: 1-1-1995
   Modifications
      Date        Description
      none        none
--------------------------------------------------------------*/


#include <stdio.h>

void main(void)
{
    float    number,    /* Stores the user's input */
             average,   /* Stores the average */
             total=0;   /* Stores the running total */
    int      count=0;   /* Counts the number of numbers */

    /* Get the first number */
    printf("Enter the next data value (0 to quit data entry): ");
    scanf("%f", &number);

    /* Calculate the total of all numbers and count the numbers */
    while (number != 0.0){
      total = total + number;
      count = count + 1;
      printf("Enter the next data value (0 to quit data entry): ");
      scanf("%f", &number);
    } /* End while */

    /* Calculate and print the average */
    average = total/count;
    printf("The average of your data is %f.\n", average);
}
```

The first lines of this program are surrounded by the symbols /* and */. Any text between these symbols is called a *comment* and is ignored by the *C* compiler. Comments allow the programmer to add some explanatory remarks into the program to make it more readable by humans. The comments at the top of the program provide some preliminary information about the program, and the comments interspersed throughout the program serve to clarify the various parts of the program.

The next statement

```c
#include <stdio.h>
```

tells the compiler to make use of some other code that is in a file called `stdio.h`. This file is part of a library of utilities for reading and printing information. The `printf()` and `scanf()` commands used in this program are part of this library.

A *C* program consists of at least one function named `main()`. A *function* is a collection of *C* statements that work together to perform some task. The line

```
void main(void)
```

is called the function header or function signature. The *function header* gives the information the function needs and the information the function returns. The two occurrences of `void` in this example indicate that `main()` does not need any information and does not return any information. In this example, `main()` contains all the statements necessary to read, calculate, and print the average of a collection of numbers.

One of the syntactical symbols that plays a major role in a *C* program is the curly brace ({). This symbol marks the beginning of a block of code, whereas its counterpart (}) marks the end of a block of code. A *block of code* is a collection of program statements that logically belong together. There are many reasons to create a block of code. In the averaging example, the outermost block of code forms the body of the `main()` function. The *function body* is where its statements reside.

Another symbol used repeatedly in a *C* program is the semicolon (;). It appears at the end of many of the lines of program code. This symbol is used by *C* to denote the end of a statement. You might think of it as a punctuation mark, like a period in English.

Once the program has been written and stored in a file, we may compile it and run it. There are many different *C* compilers and many different operating systems. Because they each behave differently and require a different set of commands, you will need to learn the specifics of yours from your manual or from your instructor.

Step 5 in the problem-solving process verifies and tests the solution. When executed, the preceding program produces the following output given the input 2, 20, 14, 5, 9, 3, and 0:

```
Enter the next data value (0 to quit data entry): 2
Enter the next data value (0 to quit data entry): 20
Enter the next data value (0 to quit data entry): 14
Enter the next data value (0 to quit data entry): 5
Enter the next data value (0 to quit data entry): 9
Enter the next data value (0 to quit data entry): 3
Enter the next data value (0 to quit data entry): 0
The average of your data is 8.833333.
```

Checking on a hand-held calculator reveals that this is the correct result. For more rigorous testing, it would be useful to run the program with real numbers, very large numbers, and very small positive real numbers. It is important to test a program very thoroughly before releasing it to the user.

Try It Using your compiler, compile the averaging program, and run it using the input data shown in step 5. Does it work correctly? Now run the program using a different set of values. Check the results on a calculator. Are they correct?

This five-step problem-solving process is demonstrated throughout this text; it is used to solve a variety of engineering programming problems. The disciplines of the applications are as follows.

Applications	Across the Disciplines	
Applications	**Discipline**	**Chapter**
Metric conversion	Civil engineering	2
Shuttle trajectory	Aerospace engineering	3
Fluid velocity	Civil engineering	4
Identifying military aircraft	Aerospace engineering	5
Simulating digital circuits	Electrical and computer engineering	6
Acid rain measurement	Environmental engineering	4, 7
Frequency distribution graphing	Industrial engineering	7

SUMMARY

This chapter has briefly introduced ANSI *C* as a programming language that can be used to solve many types of open-ended programming problems. Programming is primarily a medium for solving problems and as such should be practiced in conjunction with a sound problem-solving process. This chapter has presented just such a process. Moreover, it has shown that the five-step process can be applied to nonprogramming problems as well as programming problems. The problem-solving methodology includes the following steps:

1. Define the problem.
2. Gather information.
3. Generate and evaluate potential solutions.
4. Refine and implement a solution.
5. Verify and test the solution.

This chapter has demonstrated the application of this method on a simple problem and a simple program written in *C*. Future chapters will build on the concepts presented here.

Key Words

algorithm	high-level language
ANSI	library
ANSI *C*	machine language
block of code	object-oriented language
comment	open-ended problems
compiler	semantics
function	syntactical elements
function argument	variable
function body	while loop
function header	

2 Data Storage

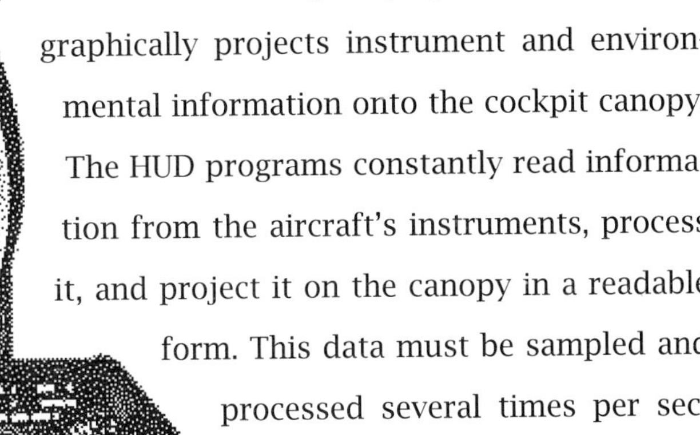

Heads-Up Display Unit Many modern military aircraft provide the pilot with a high-tech means of viewing the world from the cockpit. This device, known as a heads-up display (HUD) unit, holographically projects instrument and environmental information onto the cockpit canopy. The HUD programs constantly read information from the aircraft's instruments, process it, and project it on the canopy in a readable form. This data must be sampled and processed several times per second to provide the pilot with the most current information.

INTRODUCTION

This chapter introduces you to the methods of storing, retrieving, and manipulating data during the execution of a *C* program. All programming languages offer such capabilities. The *C* programming language offers a rich variety of data types and operators. You will learn how to use alphabetic characters to form words, use integers and real numbers to perform arithmetic calculations, and use logical data to specify true or false conditions.

In addition to the manipulation of data within a program, you will learn how to use some of the tools in the *C* library to read information from the computer's keyboard and write information to the computer's display screen. This ability to read and write information, together with the ability to manipulate data within a program, forms the essential mechanisms for the effective use of data to solve problems.

2-1 A GENERIC COMPUTER MODEL

This section presents an introductory view of the computer in order to give you a sense of how digital computers store and manipulate data.

Every digital computer is made up of five essential components: the *central processing unit (CPU),* a collection of *input devices* for providing information to the computer, a collection of *output devices* for displaying the information produced by the computer, *main memory* (also called primary memory) for immediate storage and retrieval of data, and *secondary memory* for long-term storage and retrieval of data. This collection of components is known as the computer's *hardware.*

The central processing unit is connected to each of the other four components. Figure 2-1 graphically shows the relationships among the five components of a digital computer.

To allow humans to interact with computers, computers are designed to receive input from any number of input devices such as a keyboard, mouse, joystick, touch-sensitive display, and microphone. Computers are also designed to provide output via any number of output devices such as cathode ray tube (CRT) display, liquid crystal display (LCD), printer, and speaker. The most common input and output devices are the keyboard and the computer screen or CRT display. For this reason, these devices are commonly called *standard input* and *standard output.*

The CPU is the "brain" of the computer. It is in the CPU that the instructions in a computer program are executed and the data processed. However, the CPU does not inherently know when to get data from the input devices, what to do with it, or when to send the results to an output device. For this, the CPU relies on a computer program to tell it what to do. A program is also known as computer *software.* By following the steps in a computer program, the CPU can perform its functions by interacting appropriately with the other components to help humans solve problems. To execute a program, the CPU contains the circuitry necessary to carry out arithmetic operations; very fast memory cells called *registers,* which are used to hold the data that is currently being processed; and the control units necessary to decode and execute the instructions contained in a program.

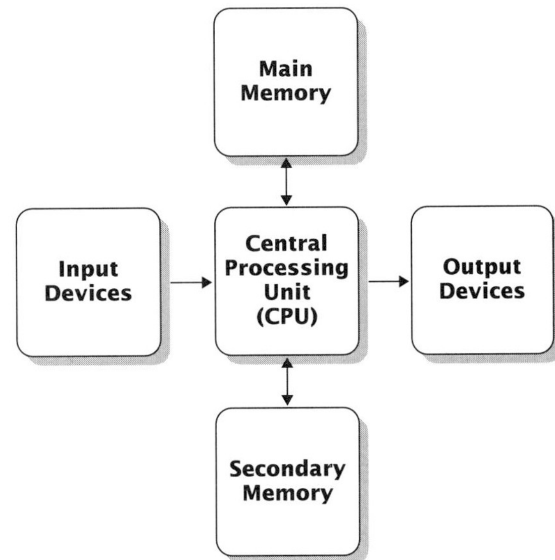

Figure 2-1 Generic Computer Model

Computer hardware and software work together to form a complete computer system.

When information is not being used by the CPU, it is stored in main memory. Main memory is *volatile,* meaning that all data is lost whenever the computer loses power. Main memory is also known as *random access memory (RAM)* because the CPU can directly access each data item in memory instead of having to search through all the data stored prior to the item of interest. A good analogue to this is the difference between a compact disc and a magnetic audiotape. The user of a compact disc (CD) player can directly access a song in the middle of a CD, whereas the user of a tape player must fast-forward through all the songs on the tape that are stored before the target song. Main memory can be thought of as a sequence of storage locations, each of which has a unique name or address by which the CPU can refer to it. Each of these locations is called a *byte* of memory. When a *C* program declares, or creates, a new data item, one or more bytes are reserved to store the new information. The number of bytes reserved depends on the type of data that is to be stored.

Because main memory is volatile, it is necessary to have a more permanent means of information storage. Secondary memory complements main memory, as it is nonvolatile. This type of memory includes floppy disk drives, tape drives, hard disk drives, optical disks, CD-ROMs, and so on. The access to this memory by the CPU is much slower than the access to main memory. Therefore, secondary memory is used to store data that is not currently being used by the CPU or is so valuable that it is worth the longer access time in order to avoid the risk of losing the data due to a power outage.

It is important to have at least an introductory understanding of how the computer works in order to be an effective programmer. Fortunately, because of high-level languages such as *C,* the programmer does not have to worry about the specific memory addresses where data is stored or where the program itself is stored. Compilers, linkers, and loaders take care of this responsibility.

2-2 SCALAR DATA TYPES

The *data type* determines how data is stored in memory as well as the legal operations that can be performed on items of that type. For instance, although it makes sense to add two numbers together, it makes little sense to add two letters together. The addition operation is typically valid only for numeric data.

Data falls into two very broad categories: composite and scalar. *Composite data* refers to information that is made up of a collection of other meaningful data items, whereas *scalar data* cannot be divided into meaningful subparts. Table 2-1 shows several examples of composite data and scalar data. This chapter will focus exclusively on scalar data types in *C*, Chapter 7 will introduce composite data types.

Scalar data types in *C* are further divided into two categories: *integral data types* and *floating point types.* Integral data types include whole numbers and characters, whereas floating point types include real numbers. The primary difference between the two categories has to do with the method the CPU uses to store the data in memory cells. This text will examine these differences in a cursory manner because much of the actual mechanics of data storage depends on the type of computer being used.

Integral Data Storage

Integral data refers to whole numbers and data that does not have a fractional part, whereas *floating point data* refers to real numbers. It is necessary for computer and electrical engineers to understand the details of how both kinds of data are digitally represented in the computer's memory.

When integral data is stored in a memory location, it is converted to its binary, or base-2, equivalent, which means that each digit in the value is either a 1 or a 0. In the same manner that base-10 place values are described as 10^0, 10^1, 10^2, and so on (from right to left), binary place values are described as 2^0, 2^1, 2^2, or ones, twos, fours, eights, sixteens, thirty-twos, and so on. A byte is a collection of *bits* or binary place values. Most commonly, a byte contains 8 bits. The leftmost bit in a byte might represent the 2^7 (128s) place in a binary number. To represent value 98_{10} (read 98 in base 10), a 1

Table 2-1   Examples of Scalar vs. Composite Data

Scalar Data	Composite Data
Your zip code	Your address (street number, city, state, etc.)
Your weight	A medical record
The weight of a manufactured part	A collection of times elapsed before the failure of a manufactured part
Wind velocity	A collection of weather statistics for a single reading of several instruments
The flight distance between two cities	A flight plan

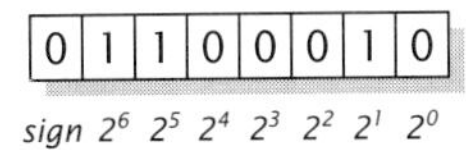

Figure 2-2 Binary Representation of 98_{10}

can be placed in the 2^6 place, the 2^5 place, and the 2^1 place, and a 0 can occupy each remaining place (see Figure 2-2). Using this method, a byte is capable of storing any base-10 value from 0 to 255 ($2^8 - 1$).

This method for encoding integer values in memory would work fine as long as it was never necessary to store negative numbers. This particular representation doesn't provide any mechanism for representing the sign of a number. To accommodate negative numbers, the leftmost bit can be used as the *sign bit*. When this bit is set to 1 the number is negative, and when this bit is set to 0 the number is positive.

With this modification, a byte is still capable of storing 256 different values, but now the numbers range from -27 to 127 ($-2^7 + 1$ to $2^7 - 1$) instead of 0 to 255. However, the sign bit scheme presents a new problem. There are two ways that 0 can be represented using this notation. Consider Figure 2-3.

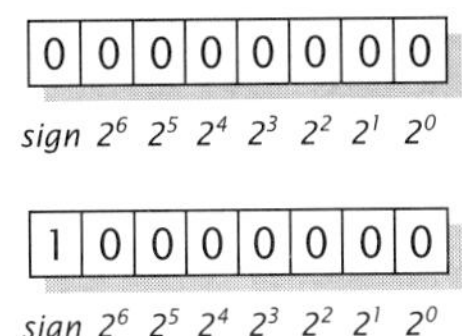

Figure 2-3 Two Ways to Represent Zero

Although -0 makes little sense, it is possible to represent -0 in this scheme. Furthermore, using this scheme, -0 is not equal to 0 because their bit patterns are different. This is certainly erroneous. Therefore, it is important that this problem be addressed.

The most common solution is known as the *twos-complement* binary representation. In this representation, the leftmost bit is still the sign bit. However, to store a negative number, the number is first stored in its positive binary form. Then all bits are "flipped" (that is, all 1s become 0s, and all 0s become 1s). Finally, 1 is added to the number. Figure 2-4 shows the conversion of -98 into its twos-complement binary representation.

Using this scheme, the number that can be stored in a single byte ranges from -128 to 127 (-2^7 to $2^7 - 1$) instead of -127 to 127. It is still possible to set the sign bit to 1 and all remaining bits to 0. (Previously, this was our erroneous "negative zero.") Now this arrangement of bits represents the number -128.

Try It

Prove to yourself that the twos-complement method eliminates the negative-zero problem.

♦ Does $-0 = +0$? Convert -0 to its twos-complement binary form by applying the conversion steps. Is the result any different from $+0$?

♦ What happened to the binary value that was -0 before the twos-complement scheme? Convert -128 to its twos-complement binary form by applying the conversion steps.

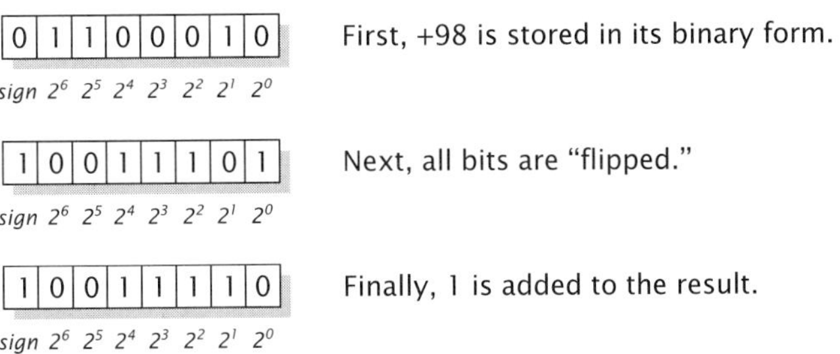

Figure 2-4 Twos-Complement Representation of -98

Integer Types in C

Fortunately, the *C* programming language hides from the programmer the details of how data is stored in memory. This means that the programmer is free to manipulate data in a form that is most familiar (for example, base 10). The *C* language supports a variety of integral data types that are based on integers (that is, they have no fractional part). These data types are called integral types and are based on the primary types: `char` and `int`. The main difference between these primary integral types is in the number of bytes reserved in memory for their storage. Most commonly, a `char` is stored in 1 byte, whereas an `int` is stored in 2 or 4 bytes, depending on the computer platform.

`char`   The `char` data type is designed for the storage of character data. Data items of type `char` are generally used to store uppercase and lowercase alphanumeric characters, punctuation characters, and special printing and nonprinting characters (spacebar, carriage return, tab, and so on.) Character constants in a *C* program are surrounded by a pair of apostrophes. For example `'x'` is a character constant. A *constant* is an actual value that appears in a program. These characters have been assigned numeric codes. This set of codes is known as the *American Standard Code for Information Interchange (ASCII)* and is widely used as a standard means of character data representation. Table 2-2 shows the base-10 ASCII values for a few of the characters in the ASCII set. See Appendix A for a complete set of ASCII codes.

The ANSI *C* standard dictates that `char` data be stored in a minimum of 1 byte. The value that can be stored in 1 byte ranges from -128 to 127. However, the standard ASCII character set consists of 128 separate characters, each associated with a value in the range 0 to 127. (No negative codes are used in the ASCII set.) In addition to the standard ASCII set, many computers support an extended ASCII set that includes special characters coded with values from 128 to 255. These characters may require the use of the leftmost bit as a placeholder for the 2^7 place rather than a sign bit. ANSI has set the minimum range for char data to be -128 to 127, even though the values from -1 to -128 are not associated with any characters.

`int`   The `int` data type is designed for the storage of integer numeric data. Data items of this type are used when it is known that real numbers are not needed. Traditionally, integer data has occupied 2 bytes of memory (16 bits) on personal computers and 4 bytes (32 bits) on workstations or larger computers. Newer personal computers store integer data in 32 bits. The ANSI standard for *C* mandates that the minimum range for `int` data must range from $-32,767$ to $32,767$ (-2^{15} to $2^{15} - 1$). This range requires 2 bytes. Computers using 4 bytes for the storage of integers are capable of $-2,147,483,648$ to $2,147,483,647$ (-2^{31} to $2^{31} - 1$).

Modifying Types

With some exceptions, programmers can modify the basic integral types `char` and `int` by preceding the basic type keyword by one of the following modifier keywords:

Table 2-2   A Sample of Some ASCII Codes

ASCII Value	Character
65	`'A'`
66	`'B'`
90	`'Z'`
97	`'a'`
98	`'b'`
122	`'z'`
48	`'0'`
49	`'1'`
57	`'0'`

- signed
- unsigned
- short
- long

These modifiers affect both the amount of memory allocated for the data storage and the meaning of each bit used in memory storage.

signed Programmers generally do not need to use the modifier signed with integral types because the default for these types is that they are signed (that is, a sign bit is used in the representation). However, in a few *C* implementations char data is unsigned by default. This modifier can therefore be used to specify that char data will be stored with a sign bit whenever you wish to be certain that the values between -128 and 127 can be stored.

unsigned This modifier allows programmers to use the sign bit of its data type as an additional place value instead of the sign bit. So, instead of an 8 bit data item of type char storing values in the range -128 to 127, it can be modified to store values in the range 0 to 255, thus incorporating the extended ASCII set. You should use this modifier only when it is certain that the data item will never need to store negative values. The range of unique values that an unsigned data item can store is no different from its signed counterpart. However, the absolute value of the maximum number is approximately twice as large when using the unsigned modifier.

short Programmers use the short modifier to conserve memory when running out of available memory is a high probability. The short modifier also ensures portability between computers that may store integers in different memory sizes (for example, 16-bit versus 32-bit machines). *Portability* refers to the ability to write a program on one computer platform and run it on other computer platforms with little or no modification. The short modifier can be used only with the int data type and will try to ensure that data is stored in 16 bits regardless of the computer's default integer size. Depending on the particular implementation, this modifier may have no effect on the storage of integral data types.

long Programmers use the long modifier when a larger range of values is required than is available with the basic int data type. Long can also be used to ensure portability. This modifier can be used only with the integral type int and the floating point type double. This modifier attempts to ensure that a data item of type int will be allocated 4 bytes for data storage. The ANSI standards for the *C* language specify that the minimum range for data of type long int is $-2{,}147{,}483{,}648$ to $2{,}147{,}483{,}647$ (-2^{31} to $2^{31} - 1$). On a computer in which the default size of an int is 32 bits, nothing is gained by specifying long int except increased portability. In fact, the increase in portability is the most common reason that programmers use the long modifier.

Floating-Point Storage

Floating-point values are also stored using a binary representation. The method of storage is significantly different from the storage of integral data because these values have a fractional part. This presents two difficult problems. To solve a broad range of arithmetic problems, real numbers need to have a high degree of precision and magnitude. *Precision* refers to the number of digits to the right of the decimal point, and *magnitude* refers to the place value of the leftmost digit in the number. The *Institute of Electrical and Electronics Engineers (IEEE)* has developed a standard for real numbers that makes use of exponential notation for the storage of floating-point values. *Exponential notation* is a method of representing a value using three parts: a mantissa, a base, and an exponent. For example, the base-10 fixed-point value 326.3851 may be represented in exponential notation as 0.3263851×10^3. In this number, the mantissa is 0.3263851, the base is 10, and the exponent is 3. The IEEE standard uses a binary exponential notation for the storage of floating-point data.

Unless you are a computer or electrical engineer, it is not important that you understand the details of this storage format. However, as a programmer, it is important to remember that there is a significant difference between integral and floating-point data as far as the computer is concerned. You should get in the good programming habit of using the correct data type for the problem being solved. If your needs require values with fractional precision, you should use floating-point data types, otherwise, use integral data types.

float   The primary floating-point data type supported by *C* is `float`. Data items of this type are guaranteed to have at least six digits of precision in their base-10 form. This data type generally occupies 32 bits for storage.

Floating-point constants in a *C* program can take several forms. First, a constant of this type can be represented as a fixed-point value with any number of digits to the left and right of the decimal point (for example, 527.17423). Second, a constant of this type can be represented as a fixed-point value with zero digits to the left of the decimal (for example, .38741). Third, a floating-point constant can be represented as a fixed-point value with zero digits to the right of the decimal (for example, 135). In this case, it is not necessary to enter the decimal point, but often 135.0 is preferable to 135 for readability. Finally, a floating-point constant can be represented in exponential form. In this case, the decimal base (10) is implied, and the mantissa is separated from the exponent by an *E* or *e* (for example, .22375E+2). It is not necessary that the decimal point be normalized to the left. Furthermore, if the sign of the exponent is omitted, it is presumed to be positive. Table 2-3 shows several examples of the use of exponential form to represent floating-point constants.

double   Often programmers need greater precision or magnitude in floating-point values. For this reason, *C* supports the data type `double`. Data items of this type have the same behaviors as items of type `float` but are typically stored in 64 bits for greater precision and magnitude. Data items of type `double` are guaranteed under ANSI guidelines to accurately represent up to 10 decimal digits of precision. For example, if you wish to ensure that the value of π is accurate to 10 decimal places, you would use a `double` instead of a float.

Table 2-3   Some Exponential Form Examples

Fixed-Point Value	Equivalent Exponential Form
3.14159	0.314159E+1
3.14159	314159.0E−5
22.375	2237.5e−2
22.375	.00022375e5
125	1.25E+2
125.0	1.25e+2

long double   Most implementations of the *C* language support the use of the modifier `long` with the data type `double`. In some cases this will extend the data storage to 124 bits for even greater precision and magnitude. However, the ANSI guidelines require only that `long double` have the same minimum precision as double, that is, 10 digits of precision. Note that `long float` would be the same as `double` and therefore is not necessary.

Table 2-4 provides a quick reference to the data types available in *C* along with their range of magnitude and their approximate storage size in memory.

Type `void`

ANSI *C* supports another type called `void`. The `void` data type is used to specify the nonexistence of a value. It may seem strange that there is a `void` data type, however, using `void` to declare certain kinds of functions makes your code easier to read. You will learn more about this type in Chapter 4.

Table 2-4   ANSI C Data Type Specifications

Data Type	Minimum Required Range	Most Common Size in Bits
char	−128 to 127	8
unsigned char	0 to 255	8
signed char	−128 to 127	8
int	−32,767 to 32,767	16 or 32
unsigned int	0 to 65,535	16 or 32
signed int	−32,767 to 32,767	16 or 32
short int	−32,767 to 32,767	16
unsigned short int	0 to 65,535	16
signed short int	−32,767 to 32,767	16
long int	−2,147,483,647 to 2,147,483,647	32
unsigned long int	0 to 4,294,967,295	32
signed long int	−2,147,483,647 to 2,147,483,647	32
float	6 digits of precision	32
double	10 digits of precision	64
long double	10 digits of precision	64 or 128

2-3 USING DATA TYPES IN A C PROGRAM

At this point, we have presented some of the technical details of data storage and the data types supported by *C*. We have not shown you how these are used in a *C* program and why they are important. Data is used in a program in many different ways and for many different purposes, which are limited only by the creativity of the programmer. In the remainder of this chapter, we show how these data types are used in programs.

Variables

The most common use of data types in *C* is in the declaration and manipulation of variables. A variable is a placeholder for a value that is used in a program. As its name suggests, the value stored in a variable can change many times during the execution of a program. Nonetheless, we must be able to refer to it. As an example, consider Example 2-1. This example contains a simplified version of the averaging program developed in Chapter 1. It calculates the average of five integers and uses the variables op1, op2, op3, op4, and op5 to hold the values to be averaged.

By the way, printf() is the mechanism used in *C* programs for printing information on the computer's screen. We will cover printf() in detail shortly.

EXAMPLE 2-1

A Program Using Variables to Calculate Averages

```
#include <stdio.h>

main()
{
  int op1, op2, op3, op4, op5; /* Variable declaration */

  /* Variable initialization: Assign initial values */
  op1 = 4;
  op2 = 9;
  op3 = 12;
  op4 = 13;
  op5 = 15;

  /* Variable manipulation: Calculate and print average */
  printf("The average of %d, %d, %d, %d and %d is: %d\n",
          op1, op2, op3, op4, op5,
          (op1+op2+op3+op4+op5)/5);
}
```

. . . .

The output from this program is

```
The average of 4, 9, 12, 13 and 15 is: 10
```

In this program the five variables op1, op2, op3, op4, and op5 were first declared, then initialized, and then manipulated to calculate their average.

These are the three important phases of the life cycle of a variable and deserve further discussion.

Variable Declaration   In a *C* program, as with many other programming languages, all variables must be declared before they are used. A variable *declaration* tells the computer to reserve some of its memory for information storage. It is important that the correct amount of memory be reserved and the appropriate storage format used for the information. For this reason, one of the aforementioned data types is used to define these characteristics while a variable name provides a reference mechanism for accessing the contents of the reserved memory.

A variable declaration in C takes the form

```
[<type modifier>] <data type> <variable list>;
```

We will use this form throughout this text to provide a general description of the syntactical elements of the *C* language. Each symbol has a special meaning. The angled brackets <> denote a nonliteral description of the item to be specified. In this syntactical description, <data type> tells us that one of the data types previously mentioned (char, int, float, double) should be specified. The square brackets [ ] indicate that the item within them is optional and that the statement is syntactically correct with or without the item. For example, [<type modifier>] tells us that the use of one of the type modifiers (signed, unsigned, long, or short) may be used but is not necessary.

The variable list takes the form

```
<variable name> [, <variable list>]
```

This a recursive definition of a variable list. A *recursive definition* is one that refers to itself in its own definition. This description tells you that a list of variables contains one variable name followed by zero or more additional variable names. In the previous example, the variable declaration was

```
int op1, op2, op3, op4, op5;
```

This variable declaration might have been modified with one or more of the type modifiers. Some examples include

```
long int op1, op2, op3, op4, op5;
unsigned long int op1, op2, op3, op4, op5;
unsigned short int op1, op2, op3, op4, op5;
unsigned int op1, op2, op3, op4, op5;
```

There are some rules as well as some guidelines for the naming of variables. In order to be a legal name, a variable must have the following characteristics.

- It must be one or more characters long.
- The first character must be a letter (a–z or A–Z).
- The remaining characters must be alphanumeric (a–z, A–Z, or 0–9) or an underscore (_).

Moreover, variable names should reflect the purpose of the variable. In the previous example the variables were simply used as operands in a mathematical expression. An operand is a value used in an arithmetic expression. Therefore, they were named op1, op2, and so on to reflect this usage. A variable that is used to store the gross income of an employee might be named `gross_salary`. Because gross income is a real number, a declaration of this data item might be:

```
float gross_salary;
```

Variable Initialization   After a declaration, a variable does not contain any meaningful information. In fact, there are no guarantees about what values are stored in variables immediately following their declaration. Therefore, it is extremely important to initialize each variable before the variable is used in any computations. Variable *initialization* simply refers to the act of storing a meaningful initial value in the variable.

Initializing Variables Using Assignment   The most common method of initializing a variable involves the *assignment operator* (=). The assignment operator is used to store actual values into a variable's memory location. In Example 2-1, the statements

```
op1 = 4;
op2 = 9;
op3 = 12;
op4 = 13;
op5 = 15;
```

use the assignment statement to initialize the variables.

The assignment operator has a variety of applications in a *C* program. We will see many different uses of this operator in future examples. However, you must know the following rules in order to use it correctly.

- A valid variable name must be on the left-hand side of the assignment operator.
- The right-hand side of the assignment operator should evaluate to the same data type as the variable on the left-hand side.

The right-hand side of an assignment expression may contain a constant (an actual value), an initialized variable, or an expression. Example 2-2 illustrates each of these uses.

EXAMPLE 2-2   Initializing a Variable Using Assignment

```
Assigning a constant to a variable
op1 = 10;
Assigning an initialized variable to a variable
op2 = op1;
Note: Since op1 was set to 10 in the previous example, op2 will
hold the same value following this instruction.
Assigning an expression to a variable
op3 = op1 + op2
```

. . . .

Using* scanf() *to Initialize Variables from User Input Another method of initializing variables is to read their values from user input. This is accomplished in a *C* program with a mechanism called scanf(). Consider Example 2-1, which printed the average of five integers. A more useful program would calculate the average of any five values. The user of the program should have the opportunity to enter those values. Example 2-3 contains a variation of the program in Example 2-1. The same five variables are used to hold the values to be averaged. In this program, the variables are initialized by using scanf() to read five values from the computer's keyboard.

<table><tr><td>EXAMPLE 2-3</td></tr></table>

Using scanf() to Average Any Five Values

```
#include <stdio.h>
main()
{
   int op1, op2, op3, op4, op5; /* Variable Declaration */

   /* Read five numbers */
   printf("Enter any five integers: ");
   scanf("%d%d%d%d%d",&op1, &op2, &op3, &op4, &op5);

   /* Print the average of the numbers */
   printf("The average of %d, %d, %d, %d and %d is: %d\n",
       op1, op2, op3, op4, op5, (op1+op2+op3+op4+op5)/5);
}
```

. . . .

scanf() is not one of the built-in *C* commands. Recall that *C* is a library-based language. *C* libraries contain functions that can be used within a program. Each function is given a name so that it can be called in appropriate parts of the program. In Example 2-3, the statement that uses scanf() is a call to the library function of that name. The scanf() function resides in a library named stdio (abbreviation for standard input and output). The standard input and output (I/O) library contains a set of functions for reading from standard input (keyboard) and writing to standard output (screen) as well as reading and writing to external files. External file I/O is discussed in Appendix B. To make use of this library, the programmer must place the following line near the top of his or her *C* program:

```
#include <stdio.h>
```

Note that this line appears at the top of the programs you have seen so far.

The scanf() function is used to read information from standard input and store it into program variables. In general, a call to the scanf() function takes the form.

```
scanf(<format control string>,<variable list>);
```

where

```
<variable list> ::= &<variable name> [, <variable list>]
```

This general form description indicates that `scanf()` takes two or more arguments. An *argument* to a function is any item that appears within the parentheses in a call to a function. In the case of `scanf()`, the first argument must be a format control string. A *format control string* is a list of format conversion codes surrounded by double quotes. A *format conversion code* provides a description of the format or data type of the incoming data. Each code corresponds to a variable in the variable list, so in a call to `scanf()` there must be one format conversion code for every variable in the variable list. A single code is represented by a percent sign followed by a letter. The letter specifies the format of the incoming data. A `%d` tells the program that a decimal integer (base 10) is being read; a `%c` denotes a character; a `%f` denotes a floating-point value; and a `%lf` denotes a double-precision float. Table 2-5 lists these `scanf()` conversion codes. There are other format codes, but these are sufficient for most applications. In Example 2-3, the statement

```
scanf("%d%d%d%d%d ",&op1, &op2, &op3, &op4, &op5);
```

contains a format control string with five codes for reading decimal integers. These codes correspond to the five variables in the variable list.

Since information is being read into the program, `scanf()` needs to know where to store this information in memory. The memory address of each variable is used to give `scanf()` this information. An ampersand (&) preceding a variable name is used by C to access the memory address of that variable. Therefore, in the statement

```
scanf("%d%d%d%d%d\n",&op1, &op2, &op3, &op4, &op5);
```

each of the variables op1, op2, op3, op4, and op5 is preceded by an &. After this call to `scanf()` is complete, these variables will contain the five values entered by the user on the computer's keyboard.

Initializing Variables in the Declaration Statement   The assignment operator can also be used directly in the declaration statement to initialize a variable. Consider the variation of the average calculator shown in Example 2-4.

Table 2-5   Format Conversion Codes and Format Modifiers

Conversion Code	scanf() meaning
%c	Reads single character input
%d	Reads integer input as a decimal value
%f	Reads floating point input in fixed point or exponential notation
%lf	Reads double precision floating point input in fixed point or exponential notation

EXAMPLE 2-4   ## Initializing Variables During Declaration

```
#include <stdio.h>

main()
{
   int op1 = 4, op2 = 9, op3 = 12, op4 = 13, op5 = 15;

   printf("The average of %d, %d, %d, %d and %d is: %d\n",
           op1, op2, op3, op4, op5, (op1+op2+op3+op4+op5)/5);
}
```

. . . .

The output of this program is the same as Example 2-1.

```
The average of 4, 9, 12, 13 and 15 is: 10
```

If used in this way, the right-hand side of each assignment operator *must* be a constant. It cannot be an expression or an initialized variable because declarations in *C* are not executable statements. Therefore, no evaluation of expressions or variables can take place in a declaration statement.

Try It Experiment with different approaches to declaring and initializing variables. Write a short *C* program that declares three variables (num1, num2, num3) designed to store *real* numbers. Write three different versions of the program that initialize the variables to 2.5, 4.23, and 3.14159 in the following ways:

♦ Initialize the variables in the declaration statement.
♦ Initialize the variables using the assignment statement, each on its own line.
♦ Read the values from the user by including the following lines:

```
printf("enter 3 real numbers and press return.\n");
scanf("%f %f %f\n",&num1, &num2, &num3);
```

♦ Finally, print the numbers using the following line:

```
printf("The numbers are: %f %f %f\n", num1, num2, num3);
```

Variable Manipulation   Once a variable has been declared and initialized it can be manipulated in many different ways. Variable *manipulation* of variables falls into one of the following categories: value modification, arithmetic expressions, or output.

Modifying the Values of a Variable   The current value of a variable can be replaced by a different value using the assignment operator. Some examples are shown in Table 2-6.

Manipulating Variables in Arithmetic Expressions   Initialized variables can be used almost anywhere a value is required. They are most commonly seen in expressions used to calculate new values or in conditional

Table 2-6   Examples of Value Modification

C Statement	Description
`var1 = 5;`	A variable can be assigned a constant value.
`avg = (5 + 3 + 8)/3;`	A variable can be assigned the result of an arithmetic expression.
`sum = var1 + 1;`	A variable can be assigned the result of an expression containing other variables.
`sum = sum + avg;`	A variable can be modified in terms of its old value.

expressions. Table 2-7 shows some examples of using variables in expressions.

Manipulating Variables using `printf()`   Initialized variables can be used to output information to standard output or to an external file. The `printf()` function is another utility that resides in the stdio library and is used to print information to standard output. In general, a call to the `printf()` function takes the form

```
printf(<format control string> [,<output list>]);
```

where

```
<output list> ::= <item> [, <output list>]
```

This general form description tells us that `printf()` takes one or more arguments. Like `scanf()`, the first argument to `printf()` is the format control string. It may contain format conversion codes that correspond to the items in the output list. Also like `scanf()`, the format conversion codes for `printf()` include %d for printing decimal integers, %c for printing characters, %f for printing floating-point values, and %lf for printing double-precision floating-point values. Additionally, the format control string may contain any number of format modifiers. A format modifier provides further control over the appearance of the output. Format modifiers appear as a backslash (\) followed by a single letter. The \n modifier will print a newline character, which moves the print position to the start of the next line on the screen. The \t modifier will move the print position one tab space (eight spaces) to the right. Table 2-8 shows these codes and modifiers. Other format conversion codes and format modifiers are supported by `printf()`, but these are sufficient for most applications. Any other characters that appear

Table 2-7   Examples of Expressions

C Statement	Description
`avg = (var1 + var2 + var3)/3;`	Operands in calculations can be variables.
`if (var4 < var3) ...`	Operands in conditional expressions can be variables.

Table 2-8   Format Conversion Codes and Format Modifiers

Conversion Code	Format Modifiers	printf() meaning
%c		Prints single character data
%d		Prints integer data in decimal form
%f		Prints floating-point data in fixed-point format
%lf		Prints double precision floating point data in fixed point format
	\n	Newline
	\t	Tab

in the format control string will simply be printed on the screen. In Example 2-1, the statement

```
printf("The average of %d, %d, %d, %d and %d is: %d\n",
       op1, op2, op3, op4, op5, (op1+op2+op3+op4+op5)/5);
```

shows a format control string containing six format conversion codes for printing decimal integers, one format modifier for printing a newline character, and several literal characters for printing a meaningful output message.

As with scanf(), each format conversion code corresponds to an output item in the remainder of the arguments to printf(). In the statement

```
printf("The average of %d, %d, %d, %d and %d is: %d\n",
       op1, op2, op3, op4, op5, (op1+op2+op3+op4+op5)/5);
```

the first five output items are simple variables, whereas the sixth is an expression that calculates the average of the five values stored in the variables.

You will see many different examples of variable manipulation throughout the remainder of this text. The variety of uses of a variable is limited only by the imagination of the programmer. It is the creative use of variables and the creative development of program logic that will largely determine your abilities as a programmer.

Try It Write a C program that will calculate 5! (five factorial), where 5! = 5 * 4 * 3 * 2 * 1. Develop the following variations of your program:

- Declare an integer variable named factorial, and initialize it with the multiplication expression 5 * 4 * 3 * 2 * 1.
- Declare an integer variable named factorial, and initialize it to 5. Write four assignment expressions that assign a new value to factorial so that the last instruction will cause factorial to contain the equivalent of 5!. The first of these expressions will be

```
factorial = factorial * 4;
```

♦ Declare an integer variable named `factorial`, and initialize it to 5 in the declaration. Write a single assignment expression that will complete the factorial equation, that is,

```
factorial * 4 * 2 * 3 * 1
```

♦ Now print the factorial of 5 by inserting the *C* instruction

```
printf("5! = %d\n", factorial);
```

In this section we introduced the life cycle of a variable. A variable's life cycle must include declaration and initialization and commonly includes manipulation. You can sometimes combine declaration and initialization into the same *C* instruction. In some circumstances, you may also be able to combine initialization and manipulation into the same instruction. This chapter has only scratched the surface of possible ways to manipulate variables in a *C* program. Throughout the remainder of this text we will introduce other possibilities.

2-4 SYMBOLIC CONSTANTS

Another principle means of using data in a *C* program is the use of symbolic constants. A *symbolic constant* is a symbol used in a program to represent a constant, or unchanging, value. Example 2-5 shows a *C* program for calculating the volume of a sphere ($4\pi r^3/3$).

EXAMPLE 2-5   ## Preliminary Sphere Volume Calculator

```
#include <stdio.h>
main()
{
   float volume, radius;

   printf("Enter the radius of a sphere: ");
   scanf("%f", &radius);

   /* Calculate the area using the formula (4*pi*R^3)/3 */
   volume = (4.0 * 3.14159 * (radius*radius*radius)) / 3.0;

   printf("The volume is: %f\n", volume);
}
```

. . . .

Given a sphere radius of 5, the output of this program is

```
Enter the radius of a sphere: 5
The volume is: 523.598328
```

This program could use some improvement, even though it does correctly calculate the volume of a sphere. Without the comment line the expression used to calculate volume is not instantly recognizable. Unless the reader of the program is familiar with π, the second operand may not be immediately obvious. Furthermore, r^3 is somewhat confusing in this expression.

The first problem can be greatly alleviated by the use of a symbolic constant. The *C* programming language supports a set of preprocessor commands that are preceded by a # symbol. The *C preprocessor* is a utility that evaluates a program just prior to compilation and acts on special commands that begin with the pound sign. We have already seen the use of one preprocessor command, namely, #include (read "pound include"). Another of these commands is #define. The #define statement is used to assign a value to a symbol in the following way:

```
#define PI 3.14159
```

Symbolic constants should be used in any case where a value is known and will remain fixed throughout the execution of the program. The names of symbolic constants are generally capitalized by convention, but this is not mandatory. The naming rules that dictate variable names also apply to symbolic constants.

All #define commands and other preprocessor commands (such as #include) are generally placed near the top of a *C* program. Although this placement is not mandatory, in most cases it is the best practice. Now consider another version of the volume program.

EXAMPLE 2-6   An Improved Sphere Volume Calculator

```c
#include <stdio.h>
#include <math.h>

#define PI 3.14159

main()
{
  float volume, radius;

  printf("Enter the radius of a sphere: ");
  scanf("%f", &radius);

  volume = (4.0 * PI * pow(radius,3)) / 3.0;

  printf("The volume is: %f\n", volume);
}
```

Given a sphere radius of 5, the output of this program is equivalent to the output of the program in Example 2-5.

```
Enter the radius of a sphere: 5
The volume is: 523.598328
```

Notice that a few minor modifications to the volume expression make it a lot more readable. In addition to the use of the symbolic constant PI, the call to the pow() function further improves readability. The built-in function pow() is defined in the math library and is designed to raise its first argument (the mantissa) to the power described by its second argument (the exponent). These improvements have negated the need for a comment explaining the equation. This programming practice is known as writing *self-documenting code* and is very desirable.

Before compiling this program, the *C* preprocessor will replace all occurrences of PI with the value 3.14159 just prior to compilation. There are numerous other applications for symbolic constants and #define statements, but this is the most common purpose.

Application 1: TRANSLATING UNITS OF MEASUREMENT

Civil Engineering

The United States is one of the few countries left in the world that does not use metric units in measurement. In order to shift to the metric system more quickly, it will be helpful to make quick translations between current measurement units and metric units. Write a program that converts distances in miles to kilometers (km) and meters (m). The conversion rate is 1 mile = 1.61 km and 1 km = 1000 m.

1. Define the Problem

We are to write a *C* program that will read a distance in miles from standard input and print the equivalent distance in kilometers and meters on standard output.

2. Gather Information

The input data for this problem is a single value that represents a distance in miles. This program will be most flexible if we allow fractions of miles to be input. Therefore, we will use data type double for the input value.

The output data will be in kilometers and meters. Because the input will be a double-precision real number, the conversion to kilometers must be of the same data type. Meters must also be of type double.

3. Generate and Evaluate Potential Solutions

The conversion from miles to kilometers is expressed by the formula km = miles × 1.61, and the conversion from kilometers to meters is expressed m = km × 1000. Multiplication in *C* is represented by * rather than x, so the *C* statements that form these formulas are

```
km = miles * 1.61;
m = km * 1000;
```

Chapter 3 will cover the arithmetic operators in *C*. The algorithm for our program is as follows:

```
Prompt for distance in miles
Read distance
Convert distance to kilometers
Convert kilometers to meters
Print distance in kilometers and meters
```

4. Refine and Implement a Solution

The following program is a direct implementation of the design expressed by the previous algorithm. Take a moment and compare the statements in the program to the steps in the algorithm. Can you see the relationship? The comments help clarify what is happening in the program. Note the use of symbolic constants to help make the program more readable.

```c
/*-------------------------------------------------------------
  Description: This program reads a distance in miles and
      converts it to kilometers and meters. Input is read
      from standard input and output is written to standard
      output.
  Programmer: Ken Collier
  Date of Last Revision: 1-1-1995
  Modifications
        Date                Description
        none                none
  --------------------------------------------------------*/

#include <stdio.h>

/* Define conversion rate as symbolic constants */
#define MILES_TO_KM 1.61
#define KM_TO_M 1000

void main(void)
{
  double  miles,    /* Initial distance in miles */
          km,       /* Distance in kilometers */
          meters;   /* Distance in meters */

  /* Prompt and read distance in miles */
  printf("Enter distance in miles: ");
  scanf("%lf", &miles);
```

```
    /* Convert miles to km and meters */
    km = miles * MILES_TO_KM;
    meters = km * KM_TO_M;

    /* Print distance in km and meters */
    printf("%lf miles = %lf km\n", miles, km);
    printf("%lf miles = %lf meters\n", miles, meters);
}
```

Given an input distance of 65, the output from this program is

```
Enter distance in miles: 65
65.000000 miles = 104.650000 km
65.000000 miles = 104650.000000 meters
```

5. Verify and Test the Solution

Checking the output from this program with a calculator reveals that it works correctly for a distance of 65. Rigorous verification requires that you hand-check a variety of outputs that include very large values, very small values, and values with several digits of precision.

SUMMARY

Programming is largely an effort in the effective management of data. The *C* programming language provides four fundamental data types—`char`, `int`, `float`, and `double`—that can help manage program data. Furthermore, *C* offers the modifiers `signed`, `unsigned`, `long`, and `short`, which provide an even greater variety of data types that programs can manipulate. It is important for programmers to ensure that each variable be declared and initialized before it is manipulated in a program.

Readability is an important characteristic of a good programming style. To enhance a program's readability, programmers should always choose the most appropriate data type when declaring variables, select meaningful variable names, and use symbolic constants instead of actual constants whenever appropriate.

Fortunately, the *C* language allows the programmer to manipulate data without requiring a detailed understanding of the internal storage and representation of the data. However, an understanding of data storage issues can help solve certain types of errors that occur as a result of the computer's limitations.

Key Words

ASCII	central processing unit (CPU)
argument	composite data
assignment operator	constant
bit	data type
byte	declaration

exponential notation
floating point type
format control string
format conversion code
hardware
IEEE
initialization
input device
integral data type
magnitude
main memory
manipulation
output device
portability
precision

preprocessor
random access memory (RAM)
recursive definition
register
scalar data
secondary memory
self-documenting code
sign bit
software
standard input
standard output
symbolic constant
twos-complement
volatile

Exercises

1. Convert the decimal value 512 into its binary equivalent.

2. How many bits are necessary to store the decimal value 512?

3. What is the largest value that can be stored in the number of bits required to store 512_{10}?

4. If the leftmost bit is reserved as a sign bit in your answer to question 3, what is the largest value that can be stored?

5. Using twos-complement arithmetic, convert the value -255 to its binary equivalent.

6. Your answer to question 5 consumes 8-place value bits plus a single sign bit for a total of 9 bits. Is it possible to store -256 in the same amount of space? Explain why or why not.

7. Is it possible to store the value 256 in the same memory configuration as -256? Explain why or why not.

8. Write a program that defines the value $\pi = 3.14159265359$ and which declares the following variables:

```
long double big_num
double med_num
float little_num.
```

   Assign the value of π to each variable and print the results. How many digits does your computer use for each data type?

9. Write a program that declares the following:

```
double num_one
float num_two
long double answer
```

   After calculating the value of answer = num_one*num_two how many accurate digits of precision would be in answer?

10. Write a program to print your name using the %c conversion code.

11. Modify the program that averages five integer values so that it will average five *real* numbers. In order for your modifications to work correctly, be sure to alter the `printf()` call so that it will print a real number instead of a decimal integer.

12. Distance traveled is calculated by the formula $D = R \times T$, where R is the rate of speed and T is the time elapsed. Design a distance calculation program in which the values for R and T are initialized to 114.56 and 78.5 respectively. Use the following shell to write your program.

```
#include <stdio.h>
void main(void)
{
        <add your solution here>
        printf("Distance travelled is %f.", distance);
}
```

   Be sure to include comments in your program. Now test your program on the computer to make sure that it works properly.

13. Modify your solution to problem 12 so that it reads the rate and time from standard input and displays the distance traveled on standard output.

14. Write a program that will read a single integer value from the user and will calculate and print the cube of that value.

15. Write a program that will read the radius of a circle as a double-precision floating-point value from the computer's standard input device (keyboard) and that will calculate and display the area of the circle on the computer's display device. Recall that the area of a circle is described as $A = \pi r^2$, where r is the circle's radius. Define the value of π (3.14159) as the symbolic constant π to make your program more readable. Use the C statement

```
scanf("%lf", &radius);
```

   to read the user's input value into a double-precision floating-point variable called `radius`.

   Use the C statement.

```
printf("%lf", area);
```

   to print the value stored in the double-precision floating point variable `area`.

16. Half-lives provide scientists with a means for dating ancient artifacts and for identifying radioisotopes. Half-life is the time required for a radioactive isotope to decay to one-half its original mass. After 1 half-life, 50 percent of the original radioisotope remains; after 2 half-lives, 25 percent remains; and so on. As an example, consider radium-226, whose half-life is 1600 years. An elapsed time of 1600 years applied to a 100-gram supply of radium-226 would result in a remaining 50 grams of radium whereas the other 50 grams would have been converted to radon plus helium.

Write a *C* program to calculate the remaining weight of a radioisotope given its initial weight, its half-life, and the amount of time elapsed since it was measured. Use the formula

$$currentweight = initialweight \times 0.5^{(timeelapsed/halflife)}$$

to calculate the current weight of a radioisotope.

Test your program on the computer using several different input combinations. Hand-calculate the current weight with the same values to be certain that your program works correctly.

17. Modify problem 14 so that the value .5 is defined as the symbolic constant HALF.

3 Computation in *C*

During the 1950s and 1960s, NASA relied on expensive, nonreusable booster rockets for its space program. By the early 1970s the booster rocket was replaced with the space shuttle—a reusable spacecraft that is launched into orbit by rockets but can return to Earth and land safely under its own power. The space shuttle consists of the orbiter, which resembles a delta-wing aircraft; the external tank, which contains oxygen and hydrogen; and the solid rocket boosters, which help the shuttle reach escape velocity before they are discarded. The space shuttle program required the collaborative effort of many engineers including mechanical, civil, electrical, aerospace, chemical, and computer engineers.

INTRODUCTION

In Chapter 2 you learned that variables and data items are an integral part of all computer programs. You discovered that the life-cycle of a variable includes declaration, initialization, and manipulation. In this chapter you will arithmetically manipulate variables using the variety of operators available in C.

The C programming language offers a rich set of operators you can use to manipulate data items. As an engineer you may need to write programs that solve engineering problems involving complicated mathematical calculations on data. To avoid errors you need to program with care, which requires a thorough understanding of the rules C uses to evaluate mathematical expressions.

3-1 RULES FOR EVALUATING EXPRESSIONS

Mathematical expressions are meaningful sequences of operators and operands. *Operators* are the symbols that describe the arithmetic action to be performed on a value or values. *Operands* are the values themselves and take the form of an actual numeric value, a variable, or a function. From previous math courses you are probably already aware that there are mathematical rules for the evaluation of a complex arithmetic expression (an expression containing multiple operators). For instance, multiplication and division operations are performed before addition and subtraction. The C language supports these rules of mathematics in its evaluation of complex expressions. It is important for you to have a thorough understanding of the manner in which C evaluates expressions. Two attributes are associated with each operator in C. The first is the operator's precedence, and the second is its associativity.

Operator Precedence

The *precedence* of an operator refers to its position in the overall operator hierarchy. Operators at the top of this hierarchy are evaluated before the operators lower in the hierarchy. Consider the expression $5 + 3 \times 4$. Knowledge of the fact that multiplication has a higher precedence than addition allows you to arrive at the correct value of this expression, 17. Without applying rules of precedence, you might have arrived erroneously at the value 32.

Operator Associativity

Many of the mathematical operators have the same precedence. It therefore is necessary to have a secondary rule for determining the order of evaluation for operations within an expression that have the same precedence. This rule is known as associativity. Operator *associativity* tells you which order to evaluate operators of equal precedence that occur within the same expression. For instance, in the expression $2 \times 3 + 10 \div 2 - 5$, multiplication and division have equal precedence, as do addition and subtraction. Because multiplication and division have higher precedence than addition and subtraction, they will be evaluated first. But which of these is evaluated

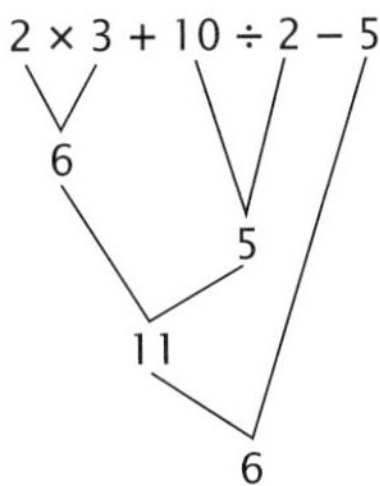

Figure 3-1 Evaluation Using Precedence and Associativity

first? These operators have a left-to-right associativity, so the multiplication will be evaluated first, followed by division. Addition will be next and, finally, subtraction. The evaluation tree in Figure 3-1 graphically represents the order of evaluation that occurs when this expression is processed.

Rules of precedence and associativity did not originate with programming languages. These are well-defined rules of mathematics that are also employed by most programming languages.

The importance of these rules becomes paramount in programming, because the computer will perform precisely the actions it is told to perform by a program. It is the programmer's responsibility to ensure that the actions in a computer program are the correct actions for solving the problem. One of the most common programming errors occurs when a programmer develops a complex arithmetic expression that is evaluated by the computer differently from how the programmer intended. A thorough understanding of these rules and the liberal use of parentheses will help the programmer avoid such pitfalls.

Table 3-1 shows the precedence hierarchy of the mathematical operators provided by *C*. The associativity of each group of operators is also provided. We will devote the remainder of this chapter to explaining and providing examples of these operators.

Try It  Apply the traditional mathematical rules of precedence and associativity to evaluate the following expressions. For the operations involved in these problems, the precedence hierarchy is

1. Exponentiation has right-to-left associativity.

2. Multiplication and division have left-to-right associativity.

Table 3-1   Some of the Mathematical Operators in *C*

Category	Subcategory	Operator	Associativity
Parentheses		()	Inside-out, left-to-right
Arithmetic	Unary	−, ++, −−	Right-to-left
	Binary first priority	*, /, %	Left-to-right
	Binary second priority	+, −	Left-to-right
Assignment		=, +=, −=, *=, /=, %=	Right-to-left

3. Addition and subtraction have left-to-right associativity.

$$2^5 + 10 \times 25 \div 5$$
$$5 + 50 \div 2^8 \times 22 - 6$$
$$5^2 \times 10 - 8 \times 2^{16} \div 3 + 19 - 4 + 5$$
$$3^{3^3} \div 2^2 \times 5^2$$

3-2 USING OPERATORS IN C

Operators in *C* fall into five general categories: parentheses, arithmetic, relational, logical, and assignment. This section discusses arithmetic and assignment operators. The relational and logical operators are introduced in Chapter 5. The operators in Table 3-1 are arranged by order of precedence, and we will discuss the operators in this order.

Arithmetic Operators

Arithmetic operators include all the operators necessary to create complex mathematical expressions. Arithmetic operators are further divided into two subcategories: unary operators and binary operators. *Unary operators* are those that are applied to a single operand, and *binary operators* are applied to two operands.

Unary Operators Unary operators have the highest precedence of all operators in *C*. Only parentheses can override their precedence. All the unary operators in *C* have a right-to-left associativity, unlike the binary operators, which have a left-to-right associativity. The unary operators provided by *C* include unary minus, increment, and decrement.

Unary minus: Unary minus is used to reflect the negativity of a numeric constant. Unary minus is a prefix operator represented by a minus sign $(-)$. A *prefix* operator is one that appears to the left of its operand(s). As an example, consider the expression 10×-2. It makes little sense to take the minus sign to mean subtraction, because it is preceded by a multiplication operator. Therefore, it is taken to mean that the second operand is the value negative 2.

Increment: The *increment* operator has the effect of incrementing the value stored in its operand by 1. The increment operator is represented by a double plus sign $(++)$ in either the prefix or postfix position. An operator is in the *postfix* position if it appears to the right of its operand(s). The operand of this *C* operator must be a single variable of an integral type. Hence, if the value stored in the variable x is 5, then following the *C* expression $++x$, its value will be 6. This operator is equivalent to the more traditional expression $x = x + 1$, in which the value stored in the variable x is given a new value that is a function of its previous value. This operator is one of the many ways that *C* provides the programmer abbreviated programming tools. Example 3-1 contains a program that uses the increment operator to count from 0 to 3.

EXAMPLE 3-1   Counting Up to 3 Using Increment

```c
#include <stdio.h>

void main(void)
{
  int num = 0;

  printf("Num is %d\n", num);
  ++num;
  printf("Num is %d\n", num);
  ++num;
  printf("Num is %d\n", num);
}
```

The output from this program looks like

```
Num is 0
Num is 1
Num is 2
Num is 3
```

Notice that this program repeats the same two statements four times. As we will see in Chapter 6, looping gives us a much better way to write a program like this. However, this example provides a simple demonstration of the use of the increment operator.

Decrement: The *decrement* operator is analogous to the increment operator except that instead of adding 1 to its operand, it subtracts 1. This operator is represented by a double minus sign ($--$) in either the prefix or postfix position. Like the increment operator, decrement requires a single integral variable as its operand. In Example 3-2, we rewrite the program in Example 3-1 so that it counts down from 3 to 0 by initializing num to 3 and using the decrement operator.

EXAMPLE 3-2   Counting Down to 0 Using Decrement

```c
#include "stdio.h"

void main(void)
{
  int num = 3;

  printf("Num is %d\n", num--);
  printf("Num is %d\n", num--);
  printf("Num is %d\n", num--);
  printf("Num is %d\n", num--);
}
```

Notice that instead of placing each decrement expression on a line by itself, we have placed it directly within the call to `printf()`. As we mentioned in Chapter 2, the arguments to `printf()` can be arithmetic expressions.

Also notice that we have placed the decrement operator in the postfix position. This example demonstrates an important feature of both the increment and decrement operators. The output from Example 3-2 looks like

```
Num is 3
Num is 2
Num is 1
Num is 0
```

If we had placed the decrement operator in the prefix position in each of the calls to `printf()`, the output would be

```
Num is 2
Num is 1
Num is 0
Num is -1
```

When the operator is in the prefix position, the value of `num` is printed after it has been decremented. When the operator is in the postfix position, the value of `num` is printed before it is decremented. In both cases the decrementation takes place. The position of the operator determines which value is used in the remainder of the statement.

In general, when you place these operators in the postfix position, the variable is altered *after* the expression or statement is evaluated. When you place these operators in the prefix position, the variable is altered *before* the rest of the expression or statement is evaluated.

In these simple examples, the effect of these operators' placement is obvious, and any errors in their placement are easily fixed. However, in a complex expression, this difference in placement of the increment and decrement operators can have a major impact on the correctness of the program. It is very important that you understand the distinction in the position of the operator. Its misuse can lead to hours of grief in tracking down the source of a program error.

Try It

What is printed by the following program?

```c
#include <stdio.h>
void main(void)
{
    int x, y=5;

    x = (--y - 10) * (240 + 6);
    printf("x = %d\n", x);
    printf("y = %d\n", y);
}
```

This next program is a very minor variation of the previous one. What is printed?

```c
#include <stdio.h>
void main(void)
{
  int x, y=5;

  x = (y-- - 10) * (240 + 6);
  printf("x = %d\n", x);
  printf("y = %d\n", y);
}
```

Binary Operators   Binary operators are divided into two groups. The first group includes multiplication, division, and modulus, and the second group includes addition and subtraction. These groupings are important because the first group has higher precedence than the second. The first group falls just below the unary operators in the precedence hierarchy and is followed in precedence by the second group. The salient details of each of these operators follows.

Multiplication: The multiplication operator multiplies its first operand by its second. Because the traditional computer keyboard does not offer a special key for multiplication, the asterisk (*) is used as the multiplication operator. It is a binary operator whose operands may be of any numeric data type. This operator has a left-to-right associativity. It performs as defined by arithmetic multiplicative laws. The *C* program in Example 3-3 provides some examples of the use of the multiplication operator.

EXAMPLE 3-3

Using the Multiplication Operator

```c
#include <stdio.h>

void main(void)
{
  int squareint;
  float squarefloat, area;

  squareint = 5 * 3;            /* Multiplying two ints */
  squarefloat = 2.5 * .00025;   /* Multiplying two floats */
  area = 3.14159 * (5 * 5);     /* Mixing ints and floats */

  printf("squareint = %d\nsquarefloat = %f\narea = %f\n",
      squareint, squarefloat, area);
}
```

The output from this program is as follows:

```
squareint = 15
squarefloat = 0.000625
area = 78.539749
```

Division: The division operator is used to divide the first operand by the second. The forward slash (/) denotes division in *C*. This binary operator may have operands of any numeric data type. Division has left-to-right associativity and performs as prescribed by arithmetic laws of division. One programming error that occurs with the use of the division operator is the divide-by-zero error. The programmer must ensure that the denominator does not evaluate to zero, or this error will occur. The program in Example 3-4 uses division to calculate and print one-half, one-quarter, and one-fifth of the value entered by the user.

EXAMPLE 3-4 Using the Division Operator

```c
#include <stdio.h>

void main(void)
{
   int number;
   float half, quarter, fifth;

   printf("Enter an integer: ");
   scanf("%d", &number);

   half = number/2.0;   /* Calculate one-half of number */
   quarter = number/4.0;   /* Calculate one-quarter of number */
   fifth = number/5.0;   /* Calculate one-fifth of number */

   printf("One half of %d is %f\n", number, half);
   printf("One quarter of %d is %f\n", number, quarter);
   printf("One fifth of %d is %f\n", number, fifth);
}
```

. . . .

Given the input value 10, the program produces the following output:

```
Enter an integer: 10
One half of 10 is 5.000000
One quarter of 10 is 2.500000
One fifth of 10 is 2.000000
```

Modulus: The modulus operator calculates the remainder of its first operand divided by its second operand. More properly known by the mathematical term *residue modulus,* the modulus operator is often referred to as the "mod" operator. The modulus operator is denoted in a *C* program by the percent sign (%). This binary operator may have operands of only an integer data type and has left-to-right associativity.

One of the easiest ways to understand this operator is to remember elementary-school mathematics. Before learning about fractions and their relationship to division, you learned to solve division problems by calculating both a quotient and a remainder. The remainder is the value expressed by the modulus operator. Therefore, the *C* statement

```
number = 23 % 5;
```

will assign the value 3 to the variable number. The program in Example 3-5 shows a common use of the modulus operator to determine whether a number is even or odd. You can take a similar approach to determine if a number is divisible by any other number simply by changing the value of the symbolic constant DENOM and rewording the input prompt.

EXAMPLE 3-5   Checking If a Number Is Even or Odd

```
#include <stdio.h>
#define DENOM 2

void main(void)
{
  int number;

  printf("Enter an integer: ");
  scanf("%d", &number);

  printf("0 means the number is even and 1 means odd: ");
  printf("%d\n", number % DENOM);
}
```

. . . .

Running this program several times using the input values 15, 10, and 7 produces the following output:

```
Enter an integer: 15
0 means the number is even and 1 means odd: 1
Enter an integer: 10
0 means the number is even and 1 means odd: 0
Enter an integer: 7
0 means the number is even and 1 means odd: 1
```

In Chapter 5 you will learn how to use the if statement to improve the way this program works.

The modulus operator offers a powerful tool for solving many different types of problems in elegant ways. Like many powerful tools, it is easy to overlook the variety of ways this operator can be used to solve problems. When faced with a difficult problem, the talented programmer will quickly inventory all the available tools in an effort to find the best one for the job.

Addition: The addition operator calculates the sum of its two operands. This operator is represented by a plus character (+). It has left-to-right associativity, and its operands may be of any numeric data type. It behaves according to the traditional rules of arithmetic addition. Example 3-6 is a program for calculating the sum of whole numbers between 0 and 5.

EXAMPLE 3-6   Calculating the Sum of Integers from 0 to 5

```c
#include <stdio.h>

void main(void)
{
  int sum;

  sum = 5 + 4 + 3 + 2 + 1;
  printf("The sum of integers between 0 and 5 is: %d\n", sum);
}
```

. . . .

The output from this program is

```
The sum of integers between 0 and 5 is: 15
```

Subtraction: The subtraction operator calculates the difference between its first and second operands. This operator is represented by a minus sign (–). It has left-to-right associativity, and its operands may be of any numeric data type. It behaves according to the traditional rules of arithmetic subtraction. Note that this operator looks the same as the unary minus operator presented previously. The context of the use of the minus sign determines whether it is interpreted as a unary or binary minus. Example 3-7 is a program that uses subtraction to calculate the distance traveled in miles given the odometer readings at the start and end of the trip.

EXAMPLE 3-7   Using Subtraction to Calculate Distance Traveled

```c
#include <stdio.h>

void main(void)
{
  float start, end, diff;

  printf("Enter initial odometer reading: ");
  scanf("%f", &start);
  printf("Enter final odometer reading: ");
  scanf("%f", &end);

  diff = end - start;

  printf("You traveled %f miles.\n", diff);
}
```

. . . .

Given inputs of 41328.5 and 44529.8, the output of this program is

```
Enter initial odometer reading: 41328.5
Enter final odometer reading: 44529.8
You traveled 3201.300781 miles.
```

Try It   Evaluate the following arithmetic *C* expressions according to the rules of precedence and associativity that are attached to each operator. What is the value of the following expressions? Assume that the variable y contains the value 5 prior to each statement.

```
5+-4*25/5%--y;
5- --y/4+10-5*5;
25*-10+500/(6%4);
--y+ ++y*10/6%4;
```

Now, place parentheses in each of the expressions to make the statement more readable without changing the value of the expression.

What If   Gas mileage is given by the formula $M = \dfrac{D}{F}$, where *D* represents the distance traveled and *F* represents the fuel consumed. Modify the program in Example 3-7 so that it also asks the user for the fuel consumption in gallons and then calculates and prints the gas mileage in addition to the distance traveled.

Assignment Operators

Although the operators presented previously are useful, they are severely limited unless we are able to store the results of arithmetic expressions into program variables. *C* offers a powerful set of assignment operators to perform this task. These operators are of the lowest precedence of the operators presented here. These binary operators each have right-to-left associativity. The rules for the use of this family of operators were presented in Chapter 2. It is important for you to remember these rules in order to avoid injecting errors into the program.

- A valid variable name must reside on the left-hand side of the assignment operator.
- The right-hand side of the assignment operator should evaluate to the same data type as the variable on the left-hand side.
- The right-hand side of an assignment expression may contain an actual or symbolic constant, a previously initialized variable, or an arithmetic expression containing any of the previous items.

Simple Assignment: You are already familiar with this operator from Chapter 2. The simple assignment operator assigns the value of its right operand to the variable that makes up its left operand. It is denoted by a single equal sign (=). This operator is the most common for assigning values to variables.

Compound Assignment: *C* supports a family of assignment operators called compound assignment operators. Compound assignment operators modify their left-hand variable by applying an operation to the value of their right operand. Incremental assignment is one member of this family.

This operator is denoted by a plus sign followed by the equal sign (+=). This operator increments the value of its left operand by the value of its right expression. Example 3-8 is a program that uses incremental assignment to add 50 to any number entered by the user.

EXAMPLE 3-8   ## Using Incremental Assignment to Add 50 to a Number

```c
#include <stdio.h>
#define CONSTANT 50

void main(void)
{
    int x;

    printf("Enter an integer: ");
    scanf("%d", &x);
    printf("%d + %d = ", x, CONSTANT);

    x += CONSTANT;

    printf("%d\n", x);
}
```

Note the use of the symbolic constant CONSTANT to make the program easier to modify for a different incremental amount. Given the input 43, the output from this program is

```
Enter an integer: 43
43 + 50 = 93
```

The assignment statement in this program could also be written using the simple assignment operator as

```
x = x + CONSTANT;
```

Compound assignment operators provide a shorthand method of modifying a variable in terms of its current value. In addition to the += operator, *C* supports the operators –=, *=, /=, and %=. Table 3-2 describes each of these operators and provides examples of their usage.

When a variable is used on both the left-hand side and right-hand side of an assignment operator, it is important to understand what is happening. Consider these rather unusual statements.

```c
int x = 5;                 /* Declare and initialize x to 5 */
x = x * ++x;
```

You must consider operator precedence when evaluating this expression. Because the assignment operator has the lowest precedence of the operators presented here, all other operations will be evaluated first. The increment operator has the highest precedence, causing x to be incremented to

Table 3-2 Compound Assignment Operators

Operator	Description	Usage	Simple Equivalent
+=	Modifies left operand by adding right operand	x += 5;	x = x + 5;
−=	Modifies left operand by subtracting right operand	x −= 5*5;	x = x − (5 * 5);
*=	Modifies left operand by multiplying it by right operand	x *= Y/4−6;	x = x * (y/4−6);
/=	Modifies left operand by dividing it by right operand	x /= 20;	x = x / 20;
%=	Modifies left operand by calculating its modulus when divided by right operand	x %= 5;	x = x % 5;

the value 6. Now, instead of the expression 5 * 5, we have the expression 6 * 6 on the right-hand side of the assignment operation. Therefore, the final value for x is 36.

Try It

Write a simple and compound assignment statement for each of the following descriptions:

♦ Set the variable `area` to hold the value of the area of a circle where the radius is stored in a variable named `radius` (πr^2).
♦ Set the variable `radius` to its current value plus 5.
♦ Set the variable `square` to its current value times its current value.
♦ Set the variable `cube` to its current value times its current value times its current value.

What If

The following code is part of a *C* program that prints a table of the areas of circles with radii varying from 5 to 25 units on increments of 5. As is, the program prints only the first row in the table. Add the necessary code so that it will print the remaining four rows.

```c
#include <stdio.h>
#define PI 3.14159

void main(void)
{
  int radius=5;

  printf("Area = %f\tRadius = %d\n", PI*radius*radius, radius);

    <your solution goes here>
}
```

Using Parentheses to Control the Evaluation

As in mathematics, programmers can use parentheses in a *C* program to override the rules of precedence and associativity that accompany each

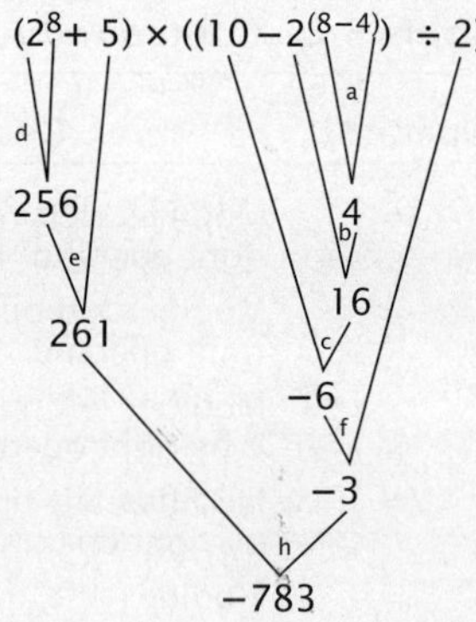

Figure 3-2 Evaluation Tree for a Parenthesized Expression

operator. Parentheses also have rules for the order of their evaluation. Parenthesized expressions may be nested within one another. Because of this nesting, parentheses have an innermost-to-outermost followed by a left-to-right associativity. When two parenthesized subexpressions are at the same nesting level, they are evaluated in a left-to-right order. For instance, in the expression $(2^8 + 5) \times ((10 - 2^{(8-4)}) \div 2)$, there are three levels of nested parentheses. Figure 3-2 shows the evaluation tree for this parenthesized expression. The order of operations is denoted by the alphabetic labeling of the branches in the tree.

What If

The following complex arithmetic expression is not particularly easy to interpret. Parenthesize this expression so that the order of operations is very clear.

$$5 - 2 + 8^2 \times -24 \times 12 - 3^5 \div 10$$

3-3 PROCESSING ARITHMETIC ERRORS

Unlike theoretical mathematics, digital computers are not capable of processing numbers that are infinitely large or infinitely small or real numbers that have infinite precision. Because a finite amount of memory is set aside for the storage of numeric data, the size or precision of that data is finite. Chapter 2 includes a discussion of data storage and provides a table of the minimum and maximum integer values. Most engineering programs involve numeric manipulation. It is therefore reasonable to consider what happens if these maximum or minimum values are exceeded by an arithmetic expression.

Avoiding Arithmetic Overflow

Arithmetic overflow refers to the effect caused by exceeding the maximum value that the computer is capable of storing in a single numeric memory location. *C* does not alert the programmer whenever arithmetic overflow occurs. Instead, the program continues executing with erroneous information. In this section we will examine why this happens.

For this discussion we will consider integer data that is stored in signed format in a single byte (8 bits). As discussed in Chapter 2, the range of values that can be stored in this space is between -128 and $+127$, inclusive. In this case, arithmetic overflow occurs anytime a value greater than $+127$ is stored in this space. The simplest example of arithmetic overflow is the addition of 1 to the value 127 (127+1). Consider this expression in binary (base-2) arithmetic. Remember that the leftmost bit is the sign bit (0 means positive, 1 means negative).

```
  0 1 1 1  1 1 1 1  (127_10)
+ 0 0 0 0  0 0 0 1  (1_10)
  1 0 0 0  0 0 0 0  (-128_10)
```

Now, using twos-complement arithmetic, we can see that this is the binary bit pattern for the decimal value -128. Therefore, using an 8-bit memory storage for integers, $127 + 1 = -128$. This is clearly an error. Just to be sure, consider adding 5 to 127. The binary expression looks like

```
  0 1 1 1  1 1 1 1  (127_10)
+ 0 0 0 0  0 1 0 1  (5_10)
= 1 0 0 0  0 1 0 0  (-124_10)
```

The result is the binary equivalent of the decimal value -124. As you can see, this problem is caused by the fact that a 1 gets carried into the sign-bit position during the addition process, which alters the sign of the value from positive to negative.

As we noted, *C* is a language that does not police the erroneous practices of the programmer. Instead it will do its best to accommodate whatever the programmer appears to want to do. If there is an error in the process, it is up to the programmer to discover the problem. Whenever your program arithmetically manipulates large numbers, and the result is a very small negative number, it signals an arithmetic overflow.

Avoiding Arithmetic Underflow

Arithmetic underflow is the counterpart of arithmetic overflow. It refers to the effect caused by referring to values that are less than the minimum value representable for a particular data type on a particular computer. Again consider an 8-bit signed integer. The minimum value that can be represented in this space is -128. Let us examine what happens when 1 is subtracted from the binary representation of -128.

```
  1 0 0 0  0 0 0 0  (-128_10)
+ 0 0 0 0  0 0 0 1  (1_10)
  0 1 1 1  1 1 1 1  (+127_10)
```

As you can see, arithmetic underflow causes a similar effect to the arithmetic overflow problem. Because of the borrowing that takes place at the sign bit, the value changes to a large positive number.

Managing Mixed-Type Expressions

Another problem that arises in programming is the mixed-type expression. A *mixed-type expression* is one whose operands are of different data types. For example, consider the program in Example 3-9.

EXAMPLE 3-9   ## Mixed-Type Expressions

```c
#include <stdio.h>

void main(void)
{
   char x=127;
   long int y=40;
   float z=3.14;

   printf("%?\n", x * y / z);   /* Which format conversion code
                                    is correct? */
}
```

The expression in the call to `printf()` contains operands of three different types. It is logical to wonder what data type the resulting value will be.

There are several approaches to handling mixed-type expressions. One possibility is that the programming language will not allow mixed-type expressions. Ada is one such language. Another possibility is that the fractional parts of the real numbers will be ignored and only the integral parts considered in evaluating the expression. The FORTRAN language behaves in this manner, which leaves the problem of whether floating-point values are rounded or truncated. *Truncation* is the effect of removing the fractional part of a real number without altering the integral part. Obviously, this approach has serious drawbacks. The final possibility is that the integer values could be converted to floating-point numbers with a fractional part of 0.0. This last approach is the one taken by *C*; it is known as automatic or implicit type conversion, coercion, promotion, or widening. For this discussion we will use the term *promotion*.

Automatic type conversion is simple to understand and follows a logical line of reasoning. In a mixed-type expression, each operation is evaluated in order using the rules of precedence and associativity, and mixed operands are promoted by applying the following rules:

1. Any operand of type `char` or `short int` is promoted to `int`.
2. Any operand of type `unsigned char` or `unsigned short` is promoted to `unsigned int`.
3. If after steps 1 and 2 the expression is still mixed, a typing hierarchy is applied to create a homogenous expression. The typing hierarchy is as follows:

int < *unsigned int* < *long int* < *unsigned long int* < *float* < *double* < *long double*

This hierarchy reflects the notion that operands of types that are lower in the hierarchy will be promoted to a type that is higher whenever necessary. So, the expression in Example 3-9 will result in a float, and the format conversion code should be %f.

You can force promotion to occur by preceding operands or subexpressions by the name of the desired type inside parentheses, as in Example 3-10. This technique is called *casting.* In this example the variable z is converted to an int and thereby loses its fractional part.

EXAMPLE 3-10   ## Using Explicit Type Conversion

```c
#include <stdio.h>

void main(void)
{
  char x=127;
  long int y=40;
  float z=3.14;

  printf("%d\n", (int)x * y / (int)z); /* Explicit conversion */
}
```

It is important to thoroughly understand mixed-type expressions. Mixed-type expressions should be avoided whenever possible through the proper declaration of variables. However, they cannot always be avoided. Earlier we advised on the liberal use of parentheses to ensure that the meaning of an expression equals the intentions of the programmer. The liberal use of casting can serve the same purpose.

Managing Uniform-Type Expressions

It is also possible for problems to arise in uniform-type expressions. Consider the program in Example 3-11.

EXAMPLE 3-11   ## Problems with Uniform-Type Expressions

```c
#include <stdio.h>

void main(void)
{
  int x=1, y=3;
  float z;

  z = x / y;
  printf("The value of z is %f\n", z);
}
```

In this case, you might think that the value 1/3 would be calculated and the result 0.33333 assigned to z. Unfortunately, this is not the case. Remember that coercion takes place as each operation is being performed. Therefore, because x and y are both of type int, they are not coerced during the evaluation of the division operation. So, the result of this division (0.33333) is coerced to an int, truncating the fractional part and returning the integral part 0. Then when the assignment operation is performed, this value is coerced to a float and z is assigned the value 0.0. The output of this program follows. Obviously, the effects of this problem are potentially disastrous.

```
The value of z is 0.000000
```

There are two solutions to this problem. The first is to declare x and y as floats, which eliminates the need for automatic promotion. This is a desirable solution whenever possible. The second solution is to manually promote the operands of the expression to float through casting. Example 3-12 reflects this modification.

EXAMPLE 3-12 Correcting the Uniform-Type Problem Using Casting

```c
#include <stdio.h>

void main(void)
{
  int x=1, y=3;
  float z;

  z = (float)x / (float)y;
  printf("The value of z is %f\n", z);
}
```

Now the value 0.33333 will be correctly stored in z as intended. The output of this program is as follows:

```
The value of z is 0.333333
```

What If  A rock initially at rest falls $16t^2$ feet in t seconds. To determine the rock's speed at a given time, you could take an interval of time, say, $t = 2$ seconds to $t = 2.01$ seconds, and determine its speed within this interval. At time $t = 2$, the rock has fallen $16(2)^2 = 64$ feet, and at time $t = 2.01$, the rock has fallen $16(2.01)^2 = 64.6416$ feet. So, in 0.01 second the rock fell 0.6416 feet. The average speed of the rock during this interval was $\frac{0.6416}{0.01} = 64.16$ feet per second. This approach provides a fairly accurate estimate of the speed at time $t = 2$ seconds. For a more accurate estimate you could shorten the interval. A more general solution to this problem should calculate the speed of the rock at time *n* using an interval of *h*. This general equation for calculating the speed of the rock from time $t = n$ to time $t = n + h$ is

$$\frac{16[(n + h)^2 - n^2]}{h}$$

feet per second.

♦ Write a *C* expression that will assign the value of this expression to the variable `feet_per_second`, where n5 is a variable of type `int` and h is a variable of type `float`.
♦ Using the automatic type conversion rules of C, what will be the resulting data type of the right-hand side of the assignment statement?
♦ Rewrite this expression using casting to ensure that the resulting type will be `double`.

3-4 THE C MATH LIBRARY

As we noted previously, *C* is known as a library-based language. So, although *C* does not have a large number of built-in functions, it is possible to extend the language by building libraries of utility functions for performing tasks that are not part of the base language but that are necessary for the creation of programs. In Chapter 4 we will show you how to build functions that might reside in such a library. Fortunately, you do not have to build all library functions yourself. The ANSI standard requires that *C* compilers be accompanied by a standard set of prebuilt libraries for use by *C* programmers. You have already seen references made to some of the *C* libraries. In Chapter 2 you discovered the stdio library and two of its functions, `printf()` and `scanf()`.

Throughout this text you will be introduced to new libraries as necessary. The math library is appropriate to the topics discussed in this chapter. This library contains several function definitions for standard mathematical operations that are not handled by the basic set of *C* operators. Further, the header file `math.h` contains special definitions for some of the utilities provided in the math library. For example, the symbolic constant `M_PI` is defined in `math.h` to represent π with a precision of 20 decimal digits. To use these definitions and the math library, it is necessary to place the following statement near the top of your *C* program:

```
#include <math.h>
```

This statement tells the *C* compiler to recognize any cails made to functions defined in the math library. Depending on the *C* compiler you are using, it may also be necessary to use a special compiler command in order to include the math library. For example, if you are using the standard Unix *C* compiler *cc*, you must use the *-lm* compiler option, which tells the com-

Table 3-3   List of Commonly Used Math Functions in the *C* Math Library

Function	Description
`double cos(double)`	Returns the cosine of its argument in radians
`double exp(double)`	Returns the natural logarithm e raised to the power of the argument
`double fabs(double)`	Returns the absolute value of its argument
`double log(double)`	Returns the natural logarithm for its argument
`double log10(double)`	Returns the base-10 logarithm for its argument
`double pow(double, double)`	Returns the value of the first argument raised to the power described by the second argument
`double sin(double)`	Returns the sin of its argument in radians
`double sqrt(double)`	Returns the square root of its argument
`double tan(double)`	Returns the tangent of its argument in radians

piler that the math library is to be used. See your compiler's manual for the details specific to your compiler.

The math library provides many useful functions. Although far from a comprehensive list, Table 3-3 outlines some of the functions most commonly used by engineers. The table lists the prototype of each function, followed by a description of the function's behavior. A function prototype is a description of the information that a function needs in order to do its job and the information that a function returns. A prototype is of the form

```
<return type> <function name> (<argument types>)
```

The prototype tells you the name of the function as well as the data type of the value that is returned by the function and the data type of any arguments expected to be passed to the function. We will cover function prototypes in greater detail in Chapter 4.

Although each of these functions returns a value of type `double` and expects arguments of type `double`, *C*'s automatic type conversion will handle other data types. In Example 2-6 you examined a program that makes use of the `pow()` function from within the math library. In this program the line

```
volume = (4 * PI * pow(radius,3))/ 3;
```

is used to calculate the volume of a sphere given its radius. This statement could have made use of `M_PI` rather than `PI` since it includes `math.h`.

Notice that the call to the `pow()` function in this statement is using one argument of type `float` (`radius`) and another of type `int` (`3`). These will each be coerced into `double` within the `pow()` function.

Another interesting thing about the use of `pow()` in this program is that it is embedded within a complex expression as if it were the operand to the multiplication operator. When you consider that this function returns a single value, it makes sense to think of functions such as this being used as operands. The fact that *C* allows programmers to use function calls in this way makes the language much more powerful and flexible.

Try It

The area of a triangle of the form

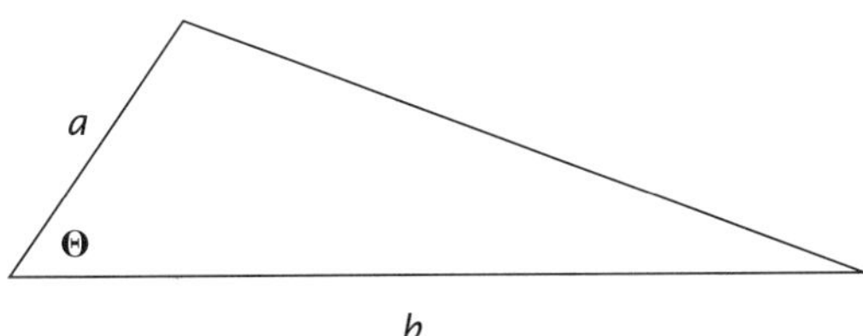

where Θ is measured in degrees, is $\dfrac{ab \sin \Theta}{2}$. The formula for converting degrees to radians is $\dfrac{\pi \Theta}{180}$.

- ◆ Write the *C* expression for converting degrees into radians. Store the result in a variable called `angle_in_radians`.
- ◆ Write a *C* expression for calculating the area of a triangle given values for the variables `a`, `b`, and `angle_in_radians`. Make sure to use casting if necessary to ensure that no information will be lost during processing.

Application 1   SHUTTLE TRAJECTORY

Aerospace Engineering

The efforts of many engineers must be coordinated to successfully launch a space shuttle into the Earth's outer atmosphere. Additionally, many calculations must be performed in order to carefully predict and track the flight of the shuttle. Computers, both on board the shuttle and at mission control, have been programmed to handle most of these calculations very rapidly so that the engineers and scientists have immediate feedback on flight data. Two data points that must be tracked continually are the vertical distance (height above the ground) and the horizontal distance (distance traveled "sideways"). The vertical distance is calculated by the formula $Y = V_y t - 16t^2$, where V_y is the initial velocity in the vertical direction and t is the time elapsed since takeoff. The horizontal distance is calculated by the formula $X = V_x t$, where V_x is the initial velocity in the horizontal direction.

Given values for t, V_x, and V_y, we can write a simple program to calculate the vertical and horizontal distances as the shuttle leaves the Earth. We will use the five-step process while solving this problem.

1. Define the Problem

The problem is relatively well defined by the preceding statement. You should write a program that will

1. Read values for t, V_x, and V_y.
2. Calculate and display the values for X and Y.

2. Gather Information

The input data for this problem will be the values for t, V_x, and V_y. These values will be in floating-point format. The units of measure will be as follows: t will be in seconds, and V_x and V_y will be in feet per second. For this example we will read these inputs from the computer keyboard. However, in a real application, these inputs might come from the computers on board the shuttle or from other computers at mission control.

The output data for this program will be X and Y as floating-point values. This data will be displayed on the computer's display monitor and will represent distance in feet.

3. Generate and Evaluate Potential Solutions

The following algorithm shows the steps necessary for the calculation of X and Y:

```
Read t
Read Vx
Read Vy
Calculate X
Calculate Y
Display X
Display Y
```

4. Refine and Implement a Solution

The following C program represents a solution to this problem. Note that the data type `double` was chosen for each of the variables except `time`. For the most typical calculations, `float` will be sufficient. However, allowing the velocities and distance to occupy more memory space provides greater latitude in the use of the program. It also offers the opportunity to demonstrate a mixed-type expression.

```c
/*-----------------------------------------------------------------
   Description: This program calculates the horizontal and
      vertical distance traveled by a space shuttle during launch
      at time t, given initial vertical and horizontal velocities
      in feet per second. The program uses the formulas:

          VertDistance = InitVertVelocity(Time) - 16(time)^2

          HorizontalDistance = InitHorizVelocity(Time)

   Programmer: Ken Collier
   Date of Last Revision: 1-1-1995
   Modifications
         Date                Description
         none                none
-----------------------------------------------------------------*/
#include <stdio.h>
#include <math.h>

void main(void)
{
  float    time;                  /* Time elapsed since lift-off */
  double   initial_vert_velo,     /* Initial vertical velocity */
           initial_hor_velo,      /* Initial horizontal velocity */
           vertical_distance,     /* Vertical distance traveled */
           horizontal_distance;/* Horizontal distance traveled */

  /* Read time, initial horizontal and vertical velocity */
  printf("Enter time since lift-off (in seconds): ");
  scanf("%f", &time);
  printf("Enter initial vertical velocity (feet/sec): ");
  scanf("%lf", &initial_vert_velo);
  printf("Enter initial horizontal velocity (feet/sec): ");
  scanf("%lf", &initial_hor_velo);

  /* Calculate vertical and horizontal distance */
  vertical_distance = (initial_vert_velo * time) -
              (16 * pow(time,2));
  horizontal_distance = initial_hor_velo * time;

  /* Display vertical and horizontal distance */
  printf("\tVertical Distance: %lf feet\n",
          vertical_distance);
  printf("\tHorizontal Distance: %lf feet\n",
          horizontal_distance);
}
```

The comments at the top of the program, called the program prologue, provide useful preliminary information to others who read the program. Together with in-line comments and meaningful variable names, it makes the program much more easily understood by others.

The output from a single execution of this program would look like

```
Enter time since lift-off (in seconds): 4
Enter initial vertical velocity (feet/sec): 200
Enter initial horizontal velocity (feet/sec):200
        Vertical Distance: 544.000000 feet
        Horizontal Distance: 800.000000 feet
```

5. Verify and Test the Solution

Hand-calculations using the input values given show that this program works correctly for these values. However, to be thorough, it is important to test this program with a broader variety of values. As an exercise, compile and run this program with the input sets {15.0, 5000, .00025}, {10.0, 0.35, 100}, and {0.5, 10000.0, 100.0}. Use a calculator to hand-check the results.

SUMMARY

Engineering programming most commonly involves the encoding of complex arithmetic expressions to solve numerical problems. *C* is a programming language that supports a rich set of built-in operators for performing these tasks. To make use of this set of operators, the programmer must develop a comprehensive understanding of the operators that are available. Furthermore, without a thorough understanding of the behavior of *C* during the evaluation of complex arithmetic expressions, it is easy for the programmer to allow errors to creep into the program. In this chapter you were introduced to the use of parentheses, arithmetic operators, and assignment operators in creating mathematical expressions in a *C* program.

Although *C* is very rich in its built-in operators, there are some potential pitfalls to avoid when writing mathematical expressions in a program. We have examined the effects of arithmetic overflow, underflow, and *C*'s rules for promotion in a mixed-type expression. Attention to the value boundaries of each data type as well as the liberal use of parentheses and casting will help you avoid these problems.

You were also introduced to the math library provided with ANSI *C* compilers. Many functions in this library can be used to solve engineering problems. These functions can serve to greatly increase your programming productivity. With an understanding of the contents of this chapter, you now have dramatically increased your programming potential.

Key Words

arithmetic operator	casting
arithmetic overflow	decrement
arithmetic underflow	increment
associativity	mixed-type expression
automatic type conversion	operand
binary operator	operator

postfix   residue modulus
precedence   truncation
prefix   unary operator
promotion

Exercises

1. Write the *C* expression for the equation

$$\text{result} = \frac{\tan A + \tan B}{1 - \tan A \tan B}$$

   What should the data type of `result` be in order to preserve the value of the expression?

2. The equation for calculating the volume of a right circular cone is

$$\text{volume} = \frac{\pi r^2 h}{3}$$

   where *h* is the height of the cone, and *r* is the radius of the base. Write this expression using *C* syntax and operators. Write the correct declarations for the variables h and r.

3. The area of an equilateral triangle is calculated by the equation

$$\text{area} = \frac{\sqrt{3}\,\text{length}^2}{4}$$

   where *length* is the length of each side. Write this expression using *C* syntax and operators. Write the correct declarations for the variables area and length.

4. The expression

$$\frac{1}{a(n + 1)}\,(ax + b)^{n+1}$$

   is the antiderivative of $\int (ax + b)^n dx$ for $n \neq 1$. Write this antiderivative as a *C* expression using the proper syntax and operators.

5. Using the formulas for calculating the vertical and horizontal distance of a launched space shuttle, we can simultaneously solve both equations to obtain a formula for *Y* in terms of *X*. To do this we can solve for *t* and substitute this value into the equation for *Y* to obtain

$$Y = \left(\frac{V_y}{V_x}\right) X - 16 \left(\frac{X}{V_x}\right)^2$$

   Write this formula as a *C* expression using proper syntax and operators. Also write the *C* statement for the declaration of all variables in the expression.

6. Suppose water costs $0.55 per gallon and that a surcharge of 10 percent is added to the bill. Although the city also assesses a utility tax of 4

percent; the surcharge is not taxed and the tax is not factored into the surcharge.

Complete the first three problem-solving steps for a program that will read the number of gallons used during a given month and will calculate the amount to be billed for that month.

7. Complete steps 4 and 5 of the problem-solving process for problem 6. Use symbolic constants for tax rate, water rate, and surcharge so that the program can easily be modified.

8. Earlier in this chapter you wrote a program to calculate gas mileage using the formula

$$M = \frac{D}{F}$$

Design and write a similar program that will modify your solution to problem 8 so that the fuel consumption is also entered by the user and gas mileage is calculated and printed.

9. Reynold's number is used to determine the friction factor of fluid traveling through a pipe. The formula for Reynold's number is

$$R = \frac{\rho v D}{\mu}$$

where ρ = fluid density, v = flow rate, D = pipe diameter, and μ = fluid viscosity.

Write a C program that reads ρ, v, D, and μ from standard input into the variables `fluid_dens`, `flow_rate`, `pipe_diam`, and `fluid_visc` and prints out Reynold's number on standard output.

10. Scientists have estimated that the blood flow through the human aorta (the main artery exiting the heart) is described by the formula $V = 2500\pi r^4$, where V is the volume of blood in cubic meters per second (m^3/sec). A healthy adult aorta has a radius of approximately 0.01 meter. Any constriction in the size of the aorta has a dramatic effect on the flow of blood from the heart to the rest of the body. A 33 percent reduction in the radius of the aorta causes an 80 percent reduction in blood flow. Doctors and medical researchers are interested in determining the percentage of reduction in blood flow for heart patients.

Using the problem-solving process, design and implement a C program that will read the radius of the aorta and will display the volume of blood flowing through the aorta in m^3/sec. Be sure to test your program thoroughly.

11. Modify your design of the solution to problem 10 so that it displays the flow-reduction percentage as compared to the average, healthy aorta.

12. The shuttle launch example presented in this chapter is somewhat artificial in that it does not consider continued acceleration. Because an ascending space shuttle accelerates out of the Earth's atmosphere, it would be better to calculate vertical distance as well as current velocity.

The following formulas account for acceleration to calculate both velocities:

$$h_t = 0.5at^2 + v_0t + h_0$$
$$v_t = at + v_0$$

where

h_t is the height at time t
v_t is the velocity at time t
a is the rate of acceleration
v_0 is the velocity at time 0
h_0 is the shuttle's initial height

Use the five-step problem-solving process to write a *C* program that will read values for a, t, v_0, and h_0 from the user; calculate h_t and v_t; and display these values on the computer's display monitor. Be sure to use meaningful variable names.

13. The Fahrenheit (F) temperature scale uses the arbitrary values of 32° and 212° to represent freezing and boiling points, respectively. The Swedish astronomer Anders Celsius (1701–1744) developed a more practical temperature-measurement scale for scientific use. In the Celsius scale (C), the freezing point and boiling point are represented by 0° and 100°, respectively. The relationship between Celsius and Fahrenheit temperatures, t_f, is expressed by the formula

$$t_f = \frac{9}{5}t_c + 32$$

    Apply the five-step problem-solving process to develop a *C* program that will read a temperature in Fahrenheit from the computer's keyboard and will display its equivalent temperature in Celsius.

14. Another temperature scale is the Kelvin scale (K), developed by the physicist William Thomson, First Baron of Kelvin. In this scale, 0°K represents "absolute zero," the temperature at which all molecular movement ceases. The equivalent Celsius and Fahrenheit temperatures are 0°K = (−273.15)°C and 0°K = −459.67°F. Each Kelvin degree corresponds to one Celsius degree. Therefore, the freezing and boiling points of water in Kelvin are 273.15°K and 373.15°K, respectively. Temperatures on the Celsius scale, t_c, are related to temperatures on the Kelvin scale t_k, by the formula $t_c = t_k - 273.15$.

    Modify your program from problem 13 so that it will also calculate and display the equivalent Kelvin temperature.

15. Newton's formula

$$T_f = (T_n - A)e^{-kt} + A$$

describes the law of cooling. T_f is the final temperature of a substance with initial temperature T_n, surrounded by a substance of temperature A, after t minutes has elapsed. In this formula, e = Euler's number

(2.71828), and k = thermal coefficient, depending on the substance being cooled.

Write a *C* program that will read the values for T_n, A, t, and k from standard input and will print T_f, nicely formatted, on standard output.

Test your program for a situation in which a 130° marble with a thermal coefficient of 0.0367 is placed in an 80° glass of water for 15 minutes. Hand-calculate your answer to ensure that your program works correctly.

16. Using the modulo operator, how would you determine if a given number was a prime number (i.e. divisible only by itself and one)? Write a program to do this.

17. The series

$$f(x) = f(a) + f'(a)(x - a) + \frac{f''(a)}{2!}(x - a)^2 + \frac{f'''(a)}{3!}(x - a)^3 + \ \ldots$$

is the general form Taylor series for $f(x)$ about $x = a$. For example, $f(x) = e^x$ can be calculated

$$e^x = 1 + x + \frac{x^2}{2!} + \frac{x^3}{3!} + \frac{x^4}{4!} + \ \ldots$$

Write a program to calculate a Taylor series for $f(x) = e^x$ to ten terms.

18. Rainfall amounts and plant irrigation requirements are often given in acre-ft. Write a program to convert from acre-ft to gallons of water given that there are 43,560 square feet in an acre and 7.48 gallons per cubic foot.

19. Flow in rivers and streams is often given in millions of gallons per day (mgd). Write a program to convert mgd to cubic meters per second.

20. If the speed of light is 186,000 miles per second, write a program to calculate the distance of a given number of light years in kilometers?

21. Write a program to convert a velocity given in meters per second to furlongs per fortnight. Note that a furlong is 220 yards and a fortnight is 14 days.

4 Modular Programming with Functions

Magnetic resonance imaging (MRI) offers an alternative to traditional X-rays or CAT scans. In an MRI session the patient lies inside a cylinder that contains a strong magnet. Radio waves are then introduced into the cylinder, causing the atoms of the body to resonate. Each type of body tissue emits characteristic signals from the nuclei of its atoms, and a computer translates these signals into a two-dimensional picture.

A complex and powerful software system processes the MRI information and generates the image. Programmers develop such complex software by dividing the system into smaller, manageable parts.

INTRODUCTION

As computer programs become larger and more complicated, it becomes difficult to think about your solution as one large program. To cope with this difficulty, most problem solvers break the problem into parts. Breaking a problem into smaller, simpler parts is a valuable technique for solving complex problems. Computer programmers have adapted this technique to programs by dividing them into manageable pieces called *modules*. C supports this modular programming approach by allowing you to divide your programs into parts called functions.

Although you have been using functions like `printf()`, `scanf()`, and `pow()`, this chapter introduces you to the details of functions. The three important aspects to modular programming are defining functions, calling functions, and passing values between functions. We will explore how to define functions in C and how to manipulate the values that are passed between functions. The contents of this chapter are crucial to your becoming a successful C programmer.

4-1 CALLING FUNCTIONS

You need to define each C function in order to provide all the necessary information about its usage. After a function has been defined, it can be used as many times as needed. Each usage of a function in a program is known as a call to that function or a *function call.* You are already familiar with function calls because you have experience making calls to functions such as `scanf()`, `printf()`, and `pow()` in previous chapters. All that is necessary to call a function in a C program is an understanding of the purpose of the function, the arguments the function expects to receive, and the information it will return. It is not necessary to understand *how* a function performs its task. In fact, one benefit of functions is that they allow you to ignore irrelevant and low-level details while solving complex problems.

In a C program, a function call may reside almost anywhere. Functions that return numeric values are often used as operands in complex arithmetic expressions. Example 2-6 illustrated this property. The following statement from Example 2-6 contains a call to the `pow()` function, which is defined in the math library.

```
volume = (4 * PI * pow(radius,3))/ 3;
```

The positioning of this function call causes the function to perform its task and return a value, and then the value returned is used as one of the operands in the expression.

Functions may or may not return a value. Functions that do not return a value generally are used as a single statement in a C program rather than embedded within a more complex expression. In general, any function call may reside anywhere in a program that a C statement is expected. This is often the case in the use of `printf()` and `scanf()`. Moreover, functions that return values may be used anywhere in a program that a value of that type may be used.

Additionally, functions are not required to accept any arguments. A function that accepts no arguments is called by simply giving its name followed by an empty set of parentheses. The stdio library contains the definition for a function called getchar(). This function provides a simple way to read a character from the computer's keyboard into a variable of type char. Example 4-1 contains a program that uses getchar() to read in a lowercase character and then converts it to its uppercase equivalent. Recall that the ASCII value of the characters *a* and *A* are 97 and 65, respectively, and that the ASCII codes of the alphabetic characters are consecutive. Furthermore, note that the difference between 97 and 65 is 32. This program simply subtracts 32 from the ASCII value of the lowercase character in order to calculate its uppercase equivalent.

EXAMPLE 4-1   ## Using getchar() to Read a Character

```c
#include <stdio.h>
#define LOWER2UPPER 32;

void main(void)
{
  char lower, upper;

  printf("Enter a lowercase character: ");

  lower = getchar();
  upper = lower - LOWER2UPPER; /* ASCII values are subtracted */

  printf("The uppercase equivalent of %c is %c\n", lower, upper);
}
```

Given the input *r*, the output of this program is

```
Enter a lowercase character: r
The uppercase equivalent of r is R
```

Notice that, although the getchar() function does not accept any arguments, the empty parentheses are still required. This syntax distinguishes a function call from a variable reference and makes the job of the compiler much simpler. In fact, you may have already noticed that all references to functions in this module are followed by an empty set of parentheses. Just as this notation makes it clear to the compiler that a function is being discussed, it also makes it clear to a textbook reader.

4-2   DEFINING FUNCTIONS

Programmers can define functions to perform any computable task. There is no limit to the complexity that a programmer can build into a function definition. As the complexity of a function increases, however, the ability to read and understand the behavior of that function decreases. Readability and understandability are two desirable characteristics of a high-quality program. For this reason, many programmers adopt the rule of thumb that a function definition should occupy no more than one printed page. This is a good habit and one we will adhere to throughout this module.

The function definition is where the algorithmic logic for performing the desired task resides. Every function must have a return type specifier, a name, a parameter list, and a body. The *return type specifier* is simply a data type that describes the type of the value to be returned from the function. The function name provides a means for calling the function. The *parameter list* provides a means of transferring arguments into the function. The body contains the actual executable statements for performing the function's task. Together, the return type specifier, function name, and parameter list form the function header or signature. The signature of a function provides all the information necessary to make calls to that function.

Defining Void Functions

Example 4-2 shows a program containing a definition for the function `my_function()` in boldface type. The function signature tells you that the return type is `void`, the name of the function is `my_function`, and the function accepts arguments of type `void`. Recall that `void` is a data type used to explicitly declare the nonexistence of a value. Therefore, `my_function()` accepts no arguments and returns no values.

EXAMPLE 4-2   ## A Simple C Function Definition

```
#include <stdio.h>

void my_function(void)
{
  printf("This is my function.\n");
}

void main(void)
{
  my_function();
}
```

. . . .

Now that the function has been defined, it may be called from anywhere within the program by simply issuing the statement

```
my_function();
```

In the preceding program, the call to `my_function()` resides in the function `main()`. Although you may not have realized it, you have been defining functions since your very first C program. Every C program must contain at least one function named `main()`. Typically, `main()` accepts no arguments and returns no values, which explains why its signature is most often

```
void main(void)
```

Of course, the program in Example 4-2 is quite pointless. A simple call to `printf()` from within `main()` would be much more clear.

Defining Functions to Return Values

You have learned that functions may be defined to return a value. As with all data, the data type of the value returned by a function must be specified. This data type specification precedes the function's name in the function definition. In general, a function definition takes the following form:

```
[<return type specifier>] <function name> ([<parameter list>])
{
     [<local variable declarations>]

     <executable statements>
     [return(<return value>);]
}
```

From this general description you can see that the return type specifier is optional. If this specifier is omitted, the return type is presumed by the compiler to be `int`, which is why the use of type `void` is important. If a function is not returning a value, the use of `void` as a return type specifier keeps the compiler from thinking that an `int` is to be returned. Additionally, the parameter list is optional. It is generally considered poor programming style to omit these options. The data type `void` should be used if a function returns no value or accepts no arguments. Local variable declarations and the return statement are also optional but do not require the `void` specification when they are omitted.

Consider the function for cubing any integer value shown in the program of Example 4-3.

The commented section at the top of the program is called a *program prologue* and provides preliminary information about the entire program. The comments that precede the `cube()` function definition are called *module prologues* and provide relevant information about each function. Together with meaningful variable names, indentation, and in-line comments, these prologues are very useful in helping others understand the structure of the program. You should adopt each of these programming habits.

The `cube()` function is defined to return a value of type `int` as specified by the return type specifier. Additionally, this function is expecting its caller to provide an argument to be manipulated. As specified in the parameter list, this argument must be of type `int` as well. Within the body of the function definition, the parameter `number` provides reference to the argument that is passed into the function. A *parameter* is a holding place for values that are passed into a function through the argument list. Furthermore, within the function body, the variable `cubed_number` is declared to be of type `int`. Variables that are needed to help perform the task of the function can be declared within the function body just as variables are declared within the body of `main()`. Finally, following the executable statements in the body of the function, the `return` statement is used to return the value stored in `cubed_number` from the function. The `return` statement may contain a single variable, a function call, or an expression. For example, we could have avoided the need for the local variable `cubed_number` by simply placing the cubing expression directly in the `return` statement.

EXAMPLE 4-3   ## The Function Definition for cube()

```
/*---------------------------------------------------------------------
   Description: This program will print the cube of any integer
      entered by the user.
   Programmer: Ken Collier
   Date of Last Revision: 1-1-1995
   Modifications
         Date                Description
         none                none
--------------------------------------------------------------------*/
#include <stdio.h>


/*---------------------------------------------------------------------
Module: cube()
Description: Calculates and returns the cube of its argument.
Input Parms: An integer to be cubed.
Returns: The integer cube of its argument.
Modifications
         Date                Description
         none                none
--------------------------------------------------------------------*/
int cube(int number)
{
  int cubed_number;         /* To store result of cubing operation */

  cubed_number = number * number * number;
  return(cubed_number);
}

void main(void)
{
  int num,                  /* Original Number */
      num_cubed;            /* Result of cubing num */

  printf("Enter any integer: ");
  scanf("%d", &num);

  num_cubed = cube(num); /* Function call */

  printf("%d^3 = %d\n", num, num_cubed);
}
```

. . . .

Given an input value of 5, the output of this program is

```
Enter any integer: 5
5^3 = 125
```

EXAMPLE 4-4   ## A Variation on the Definition of cube()

```
int cube (int number)
{
      return(number * number * number);
}
```

. . . .

It is important that the data type of the expression in the `return` statement match the data type of the return type specifier. These two components are directly related to each other. The return type specifier describes the type of information to be returned, and the `return` statement describes the value that is to be returned. This relationship is reflected in Figure 4-1.

```
#include <stdio.h>

int  cube(int number)
{
    int cubed_number; /* To store result of cubing operation */

    cubed_number = number * number * number;
    return (cubed_number);
}

void main (void)
{
    int num,            /* Original Number */
        num_cubed;      /* Result of cubing num */

    printf("Enter any integer: ");
    scanf("%d", &num);

    num_cubed = cube(num); /* Function call */

    printf("%d^3 = %d/n", num, num_cubed);
}
```

Figure 4-1 Return Type and Function Type Must Match

Try It   Write the definition for a function named `circle_area()` that will accept the radius of a circle as float value and will return the area of a circle of this radius using the formula area = πR^2. Define a simple `main()` function to test `circle_area()` using the radius 3.5.

Variable Scope

The *scope* of a variable refers to the places within a program where the variable is accessible for manipulation. Variables are either global or local in scope. A *global variable* is a variable that is accessible to the entire program, whereas a *local variable* has limited accessibility.

A *C* program is made up of one or more blocks of code. The body of a function is a block of code. As mentioned in Chapter 1, code blocks are delimited, or set apart, by an opening and closing curly brace, { and }. Blocks are used to group *C* statements for a variety of reasons. Note that the bodies of the function definitions for `cube()` and `main()` in Example 4-3 are delimited by a set of curly braces.

A global variable is a variable whose declaration does not appear within any block of code, whereas a local variable is a variable that is declared inside a block of code and is accessible only within that block. In Example 4-3, the variable `cubed_number` is local in scope. Any reference to this variable in the body of `main()` would cause an error because this variable is not accessible outside the definition of `cube()`. Similarly, the variables `num` and `num_cubed` are local to `main()` and are accessible only within the body of `main()`.

Now consider the variation of the cube-calculating program in Example 4-5. Here, the variables `cubed_number`, `num`, and `num_cubed` are declared outside of any code block. This positioning of the declaration causes these variables to be global in scope. Now these variables may be referenced from anywhere within the entire program file.

EXAMPLE 4-5   Declaring Variables Globally

```c
#include <stdio.h>

int cubed_number, num, num_cubed;

int cube (int number)                          /* Function Definition */
{
  cubed_number = number * number * number;
  return(cubed_number);
}

void main(void)
{
  printf("Enter any integer: ");
  scanf("%d", &num);

  num_cubed = cube(num);                       /* Function Call */

  printf("%d^3 = %d\n", num, num_cubed);
}
```

. . . .

You may wonder why you would declare variables locally when global declarations provide such a wide range of access. The fact that global variables have such a large scope creates programming difficulties. Suppose your program is quite large and has an error in it that causes a variable to contain an incorrect value. If this variable is global and is accessed in many different code blocks in your program, it may be very difficult to identify where in your program the error occurred. Conversely, if the variable were local to a particular function, then its limited scope greatly simplifies the debugging process. Global variables are generally considered poor programming practice and should be avoided in favor of local variables whenever possible.

Try It

Earlier you wrote a short program using your definition of `circle_area()` to calculate the area of a circle of radius 3.5. Modify that program so that all variables are global in scope. Test your new version to be sure it still works correctly.

Function Prototyping

As with program variables, you must declare a function before it can be used. One way to accomplish this is to place the entire function definition before all calls to that function in a program file. The previous examples take this approach; however, function prototypes offer a method for declaring a function before it is defined. A function prototype is analogous to a variable declaration. As noted, all the relevant information necessary to use the function is provided by the function signature. The function signature can therefore serve as a function declaration. When a function signature is used to declare a function, it is called a *function prototype.* We used function prototypes in Chapter 3 to describe some of the functions that are found in the math library. Example 4-6 shows a variation of the number-cubing program that uses a function prototype for the cube() function.

EXAMPLE 4-6   ## Using Function Prototypes in the Cube Program

```c
#include <stdio.h>

int cube (int number);              /* Function Prototype */

void main(void)
{
  int num, num_cubed;

  printf("Enter any integer: ");
  scanf("%d", &num);

  num_cubed = cube(num);            /* Function Call */

  printf("%d^3 = %d\n", num, num_cubed);
}

int cube (int number)               /* Function Definition */
{
  int cubed_number;

  cubed_number = number * number * number;
  return(cubed_number);
}
```

. . . .

Function prototypes tell the compiler that the functions are defined elsewhere in the program and that calls to these functions should be accepted. Prototypes are sometimes called *forward references* because they provide an early reference to a definition that is to appear later.

Unlike the prior examples, it is preferable to define `main()` as the first function in a program, because it is the first function called at run-time. The `main()` function should be preceded by all the function prototypes, which will help others who read the program to understand the program's organizational structure.

You may have noticed that in Example 4-6 we omitted the program and module prologues. Although these prologues help clarify the program, they also tend to clutter it visually. You can solve this problem by placing prologues, prototypes, and symbolic constants into a separate file called a *header file* and removing them from the source file. By convention, a header file has the same base name as its corresponding source file, but it is followed by a `.h` extension rather than a `.c`. If the program in Example 4-6 were stored in a file named `cubes.c`, the header file would be named `cubes.h`. Example 4-7 displays the contents of `cubes.h`.

EXAMPLE 4-7   A Header File for the Cubing Program

```
/*-----------------------------------------------------------------
   Description: This program will print the cube of any integer
      entered by the user.
   Programmer: Ken Collier
   Date of Last Revision: 1-1-1995
   Modifications
         Date               Description
         none               none
   ------------------------------------------------------------*/


   /*--------------------------------------------------------------
Module: cube()
Description: Calculates and returns the cube of its argument.
Input Parms: An integer to be cubed.
Returns: The integer cube of its argument.
Modifications
         Date               Description
         none               none
   ----------------------------------------------------------*/
int cube(int number);

. . . . . . . . . . . . . . . . . . . . . . . . . . . . . . . . . . .
```

In order for this file to be included during the compilation of `cubes.c`, it is necessary to place the preprocessor statement

```
#include "cubes.h"
```

near the top of the `cubes.c` file. You have been using header files in your programs when you include `stdio.h` and `math.h` in your programs. These are simply header files for preexisting libraries. Instead of delimiting the new include file with angled brackets as we have for `stdio.h`, it is delimited by double quotes. Double quotes tell the preprocessor to look for the header file in the current directory, whereas angled brackets tell the pre-processor to look in a special library directory.

By separating your program into a header file and a source file, you can make it easier for others to understand. A reader of your program can first look at the header file to develop an understanding of the general structure of your program and then look at the source file to examine the logic details of each function.

Try It

For your program that uses `circle_area()` to calculate the area of a circle of radius 3.5, modify the program so that your definition of `main()` appears before your definition of `circle_area()`. Be sure to write the prototype for `circle_area()` before `main()`. Test your new version to be sure it still works correctly.

4-3   PASSING INFORMATION VIA PARAMETERS

Thus far we have skirted the details of how information is passed from a function call to that function, though we have been using some of the important terminology in its proper context. In a mathematical expression such as $y = f(x)$ we say that x is an argument to the function f. This terminology carries over to programming as well. In a function call, the information contained within the parentheses is known as the function's *actual arguments,* and the placeholders in the function definition are called *formal parameters.* As you have seen, an actual argument can be a simple variable, an expression, or another function call as long as the evaluation of the argument results in a meaningful value.

When you define a function, it is impossible to predict the origin of the information that the function will use to perform its task. For instance, consider the `cube()` function defined in Example 4-3. Although we could have placed a call to `scanf()` directly within the definition of `cube()`, we wanted this function to be as general-purpose as possible, allowing it to receive the value to be cubed from a variety of sources. Formal parameters make this possible. With the use of parameters, the origin of the value to be cubed is irrelevant. We can simply calculate and return the cube of the parameter whether the original value was read from the computer's keyboard or was the result of another calculation within the program. Because the body of the function must refer to the number in order to perform the calculation, the parameter name provides a temporary name by which the value of the actual argument may be referred.

In essence, a formal parameter is a local variable that is defined within the parameter list rather than the function body. The primary difference is that it receives its initial value from the actual argument in the function call. In functions that take multiple parameters, the ordinal position of each actual argument determines to which parameter it corresponds. That is, the value of the first actual argument is assigned to the first formal parameter in the list, the value of the second argument to the second parameter, and so on. Note that, as with any other data item in a *C* program, the data type of the formal parameter must be specified. Moreover, the data type of the actual argument should match the specified data type of the formal parameter. If the data types do not match, the compiler is likely to issue a warning message. If the warning is ignored, *C* will attempt to convert the value of the actual argument into the data type of the formal parameter. This conversion can have negative side-effects and should be avoided.

So far you have used formal parameters only for passing values *into* a function. You could pass values back from a function only via the `return` statement. This raises an interesting question. What happens if you change the value of a formal parameter within the body of a function? What effect will it have on the actual argument? Consider Example 4-8, which is another version of the `cube()` function seen earlier. In this version, the formal parameter `number` is being assigned its original value times its original value squared.

EXAMPLE 4-8

Changing the Value of a Formal Parameter in a Function

```
#include <stdio.h>

int cube(int number);

void main(void)
{
  int x = 3, y;

  y = cube(x);
  printf("x = %d\ty = %d\n", x,y);
}

int cube (int number)
{
    number *= number * number;    /* Value of number is altered */
    return(number);
}
```

. . . .

In `main()`, the variable x is passed as an actual argument to `cube()`, the return value is stored in y, and the values of both variables are then printed. What will be printed by the call to `printf()`? It should be fairly clear that y will contain the value 27. A more difficult question is whether x will contain its original value, 3, or the final value of the formal parameter `number`, 27. On the one hand, it seems that local changes to a formal parameter should have no external effect on the actual argument because doing so could cause serious side effects. On the other hand, because the actual argument and the formal parameter are logically connected to each other, it would make sense that changes to the formal parameter should also affect the actual argument. In fact, both possibilities are reasonable, and many programming languages support both. The first approach is a parameter-passing mode known as pass-by-value, and the second is known as pass-by-reference.

Try It   Write the definition for a function called `my_putchar()` that will accept a single character argument and print it to standard output at column 40 of the current line using `printf()`. Define a simple `main()` function to test `my_putchar()` by sending the character literal x.

Parameter Passing by Value

By default, *C* uses the pass-by-value parameter-passing mode. *Pass-by-value* means that only the value of the actual argument is passed into the function, and any changes made to the formal parameter will have no effect on the actual argument. So, the output of Example 4-8 is

$$x = 3 \quad y = 27$$

Pass-by-value has the effect of allowing the actual argument to retain its value no matter what changes are made to the corresponding formal parameter inside the body of a function. This parameter-passing mode is the safest because the programmer does not have to worry about the potential side-effects of changing the value of a formal parameter within a function. A *side effect* occurs when a change to a data item in one part of the program causes another part of the program to behave differently.

Let's examine what happens during a call to a function. Recall that a variable is simply a reference to a memory location containing a value. A formal parameter is no different in this respect. When a function is called, a memory location is allocated for each formal parameter in the parameter list. In pass-by-value mode, the value stored in the actual argument's memory location is copied into the corresponding formal parameter's memory location. Because the formal parameter contains only a copy of the value, changes made to the formal parameter affect only the copy — they have no effect on the original. Figure 4-2 provides a graphical depiction of this concept using the program in Example 4-8.

In this diagram each memory location is represented by a box. The number below the memory location represents the address of that location, and the letter to the left of the box represents the variable that is associated with

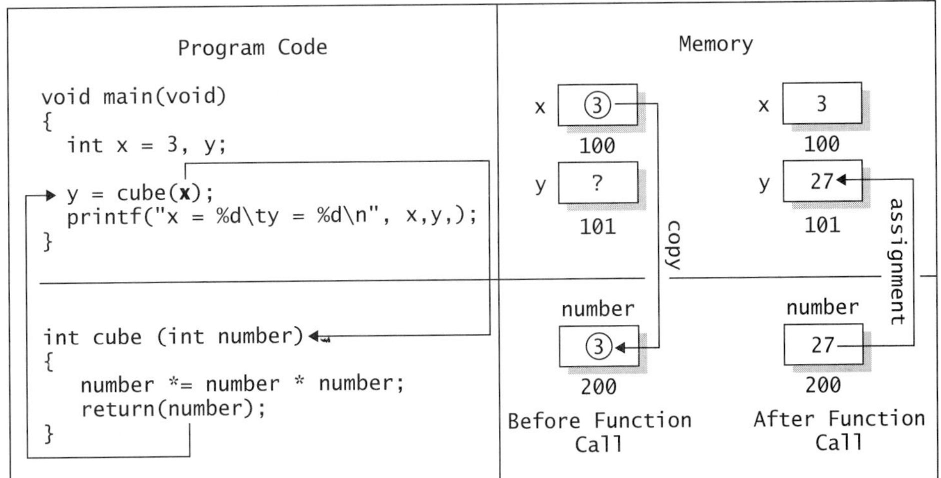

Figure 4-2 Graphical Representation of Pass-by-Value

the memory location. Notice that changes made to memory location 200 (number) have no effect on the contents of memory location 100 (x). The only way that the cubed value of x is captured in main() is by the use of an assignment operation to store the return value of cube() into y.

Parameter Passing by Reference

The converse of pass-by-value is pass-by-reference. *Pass-by-reference* means that a reference to the memory location of the actual argument is passed into the function, and any change made to the formal parameter *will* effect the actual argument. The reference that is passed into the function is the address of the actual argument's memory location. So, instead of changing a copy of the value of the actual argument by altering the parameter, the original value itself is changed. Figure 4-3 graphically depicts the behavior of the program in Example 4-8 as if *C* were a pass-by-reference language. Because number now refers to memory address 100, any changes made to number affect the contents of that memory location.

Although pass-by-value is generally the safest parameter-passing mode, pass-by-reference is often desirable and helps programmers solve certain problems that cannot be solved in a pass-by-value environment. One of these problems is the need to return multiple values from a function. The return statement in a *C* function can return only a single value. You have seen examples of functions that return no values and functions that return a single value. In each case, the parameter list is used only for passing information into a function. However, the only way you can return multiple values is to use the parameter list to pass information out of a function. This approach requires parameters that are passed by reference rather than value.

A simple, and classic, example of returning values via formal parameters involves the writing of a function that will swap the values stored in two variables with one another. Conceptually, swapping values is simple to understand; however, programming languages do not generally have a simple, straightforward method for swapping the values of two variables. You must create the function swap() to perform this task. The swap() function must

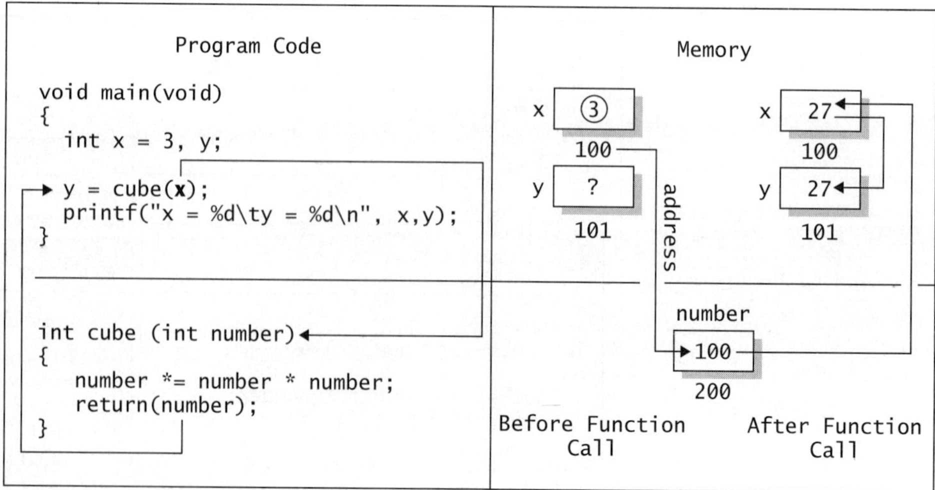

Figure 4-3 Graphical Representation of Pass-by-Reference

be designed to accept two variables, exchange their values, and return both variables.

Exchanging the values of two variables is a stimulating puzzle. Suppose the variables x and y contain the values 5 and 10, respectively, and you wish to exchange their values so that x will contain 10 and y will contain 5. If your program contains the statements

```
x = y;
y = x;
```

the original value of x will be lost after the first statement is executed, and y will never receive the value 5 as intended.

Successfully exchanging the values of two variables requires a third, temporary variable to store a copy of one variable. The algorithm for swapping two values using a temporary variable is

1. Copy the contents of the first variable into the temporary variable.
2. Copy the contents of the second variable into the first variable.
3. Copy the contents of the temporary variable (a copy of the first number) into the second variable.

Example 4-9 shows a program for swapping the values of the variables x and y using the temporary variable z.

EXAMPLE 4-9   Swapping the Values of Two Variables

```
/*-------------------------------------------------------------------
   Description: This program will exchange the values of the
       variables x and y using a temporary variable z.
   Programmer: Ken Collier
   Date of Last Revision: 1-1-1995
   Modifications
           Date              Description
           none              none
-----------------------------------------------------------------*/
#include <stdio.h>

void main(void)
{
   int x=5,              /* First variable */
       y=10,             /* Second variable */
       z;                /* Temporary variable */

   printf("x = %d\ty = %d\n", x, y);

   z = x;                /* Make a copy of x */
   x = y;                /* Overwrite x with y */
   y = z;                /* Overwrite y with copy of x */

   printf("x = %d\ty = %d\n", x, y);
}
```

The output of this program is

```
x = 5   y = 10
x = 10  y = 5
```

Because swapping values is necessary in many different applications such as sorting numbers, a function for swapping values makes sense. Using the same algorithm, Example 4-10 offers a potential definition for the function swap(). Note that because no value is being returned via a return statement, the return type specifier is void.

EXAMPLE 4-10   A Potential Definition for the swap() Function

```
void swap(int first_number, int second_number)
{
    int temp_number;

    temp_number = first_number;
    first_number = second_number;
    second_number = temp_number;
}
```

. . . .

To use this function in a program might involve a call like the following:

```
swap(low, middle);
```

This function has a problem, however. Recall that by default, *C* passes parameters by value rather than by reference. This default parameter-passing mode means that the swapping of values of formal parameters will have no effect on the values of the actual arguments. As is, the swap() function simply represents wasted processing time. Fortunately, *C* offers a means of passing parameters by reference. In fact, scanf() is just such a function, though it requires an understanding of something called a pointer in *C*.

An Introduction to Pointers in C   A *pointer* is a special type of variable that contains a memory address rather than a data value. As you have seen, most variables are intended to store a value from one of the data types in *C*. Additionally, you have seen that a variable is simply an alias for a memory address. Like all variables, pointer variables are also aliases for memory addresses; however, the information stored in that memory location is the address of another memory location. So the contents of a pointer variable refers to, or "points to," another memory location. Figure 4-4 is a graphical representation of this concept. Memory addresses are shown in italics. Because num_ptr contains the address of num, we say that num_ptr *points to* num.

Pointer variables provide two ways to access the same data value. You can access the value by referring to either its simple variable name or the

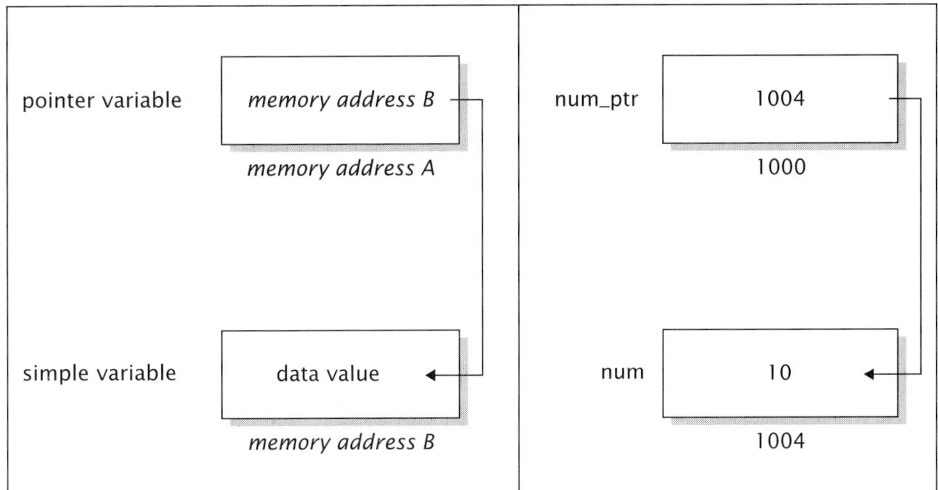

Figure 4-4 A Graphical View of Pointers

pointer name. Pointers are useful in many different and complicated ways. We will explore only the simplest of these in this module.

A pointer variable is declared almost exactly like any other variable in a *C* program. The only additional feature is that the variable name is preceded by an asterisk (*) in the declaration statement. The declaration

```
int *num_ptr, num = 10;
```

tells the compiler that the variable num_ptr is a pointer variable that will eventually point to a memory location designed to hold an int, and num is a traditional variable of type int that is initialized to the value 10.

Like all variables, num_ptr does not contain any meaningful value until it is initialized; that is to say that num_ptr is not currently pointing to any meaningful memory location. You can set num_ptr to point to num by using the address-of operator. The *address-of operator* is a unary operator whose operand must be a variable.[1] The address-of operator evaluates to the memory address of its operand. This operator is represented by an ampersand (&) in the prefix position. The following assignment statement initializes num_ptr to hold the memory address of the variable num.

```
num_ptr = &num;    /* num_ptr gets the address of num */
```

Suppose that num has been allocated memory address 1004_{10} and num_ptr the address 1000_{10}, as in Figure 4-4. Following this statement, num_ptr contains the value 1004_{10}. The contents of num can now be accessed by dereferencing num_ptr. *Dereferencing* a pointer refers to accessing the value that a pointer indicates. The asterisk is also used to dereference a pointer variable.

---

[1] For completeness, you should know that the address-of operator is of the same precedence and associativity as the other unary operators. It falls just below parentheses and has right-to-left associativity.

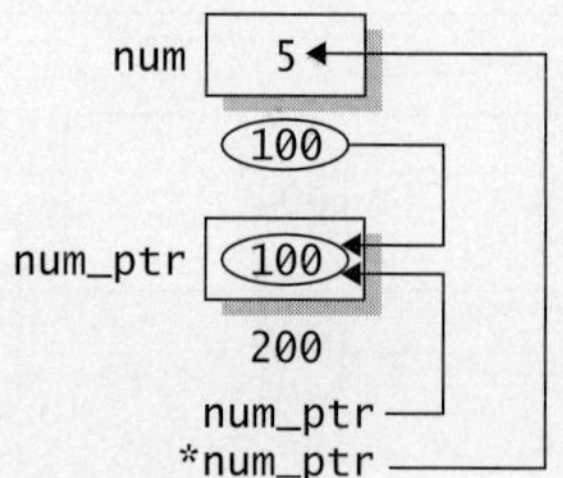

Figure 4-5 Pointer Dereferencing

Note that `num_ptr` is now associated with *three* values: its own memory address (1000_{10}), the memory address that it points to (1004_{10}), and the contents stored in that memory address (10). To print the value of `num`, you would traditionally make the following call to `printf()`:

```
printf("%d", num);              /* 10 is printed */
```

If you were to print `num_ptr` in the same manner

```
printf("%d", num_ptr);          /* 1004 is printed */
```

the value stored in `num_ptr` (memory address 1004_{10}) is printed using decimal integer format. It is not likely you are interested in this value. Instead, you must dereference `num_ptr` pointer by preceding it with an asterisk to access the value stored in memory location 1004_{10}.

```
printf("%d", *num_ptr);          /* 10 is printed */
```

The dereference expression `*num_ptr` is read "the value pointed to by `num_ptr`." A dereferenced pointer is no different from a simple variable. It can be manipulated in all the same ways and is subject to all the same restrictions, depending on its data type. So we can print the value of `num` via `num_ptr` by dereferencing it in a call to `printf()`. Figure 4-5 graphically depicts the concept of dereferencing pointers in *C*.

In Chapter 3 we introduced the asterisk as a binary operator for performing multiplication. As shown here, the asterisk is also a unary operator for declaring and dereferencing pointers. Like other unary operators, the unary asterisk is in the highest precedence class and has right-to-left associativity. Depending on the context of the asterisk within a program, the compiler interprets it as a pointer declaration, pointer dereference, or multiplication operator.

***Simulating Pass-by-Reference in* C**   You can use pointers in many creative ways to solve difficult problems. We will leave most pointer topics to be covered in an advanced textbook on *C*. But as you saw earlier in this chapter, passing parameters by reference is necessary for solving a wide variety of programming problems in a modular style. Parameter passing by reference is achieved in *C* through the use of pointers as formal parameters.

Having formal parameters that point to the actual arguments enables you to make changes to the actual values rather than to copies of the actual values, thereby simulating pass-by-reference. Rather than passing the value of a variable as an actual argument, you pass the memory address of the actual argument by using the address-of operator. Instead of declaring formal parameters like traditional variables, you declare them as pointer variables. Now, when the function call is made, the address of the actual argument is passed into the formal parameter. Within the function, whenever the formal parameter is dereferenced, access is made to the actual argument, giving the effect of pass-by-reference. Using this concept, Example 4-11 shows the correct definition of `swap()`.

EXAMPLE 4-11   ## The Correct Definition of the Function swap()

```
void swap(int *first_number, int *second_number)
{
    int temp_number;

    temp_number = *first_number;
    *first_number = *second_number;
    *second_number = temp_number;
}
```

Let's step through this definition to ensure that you fully understand the details. First, notice that the basic logic, structure, and variable names are no different from those in Example 4-10. The only difference is the use of the asterisk to declare and dereference pointer variables. If this function is passed to the addresses of the actual arguments, the formal parameters first_number and second_number will point to those arguments. Notice that the local variable temp_number is not declared as a pointer. Because this variable's purpose is to temporarily store a copy of *first_number, it may be a simple variable. In the body of swap(), all uses of first_number and second_number are dereferenced, because we wish only to change the values pointed to by those parameters. You can read the steps in the swap() function as

1. Store a copy of the value pointed to by first_number into temp_number.
2. Copy the value pointed to by second_number into the location pointed to by first_number.
3. Copy the value stored in temp_number into the location pointed to by second_number.

Using pointers, the swap() function now works as intended. It is also necessary to call the function properly. Example 4-12 shows a program that reads two integers from the keyboard, swaps them, and prints out the result. Although this is not a very stimulating program, it demonstrates the swap() function.

EXAMPLE 4-12   ## A Program that Demonstrates the swap() Function

```
/*-------------------------------------------------------------
   Description: This program will read two integers from
      standard input, swap them, and print them.
   Programmer: Ken Collier
   Date of Last Revision: 1-1-1995
   Modifications
        Date            Description
        none            none
-----------------------------------------------------------*/
```

```
#include <stdio.h>

void swap(int *first_number, int *second_number); /* Prototype */

void main(void)
{
  int first, second;

  printf("Enter two integers: ");
  scanf("%d%d", &first, &second);

  swap(&first, &second);          /* Passing memory addresses */

  printf("Your numbers swapped are %d and %d\n", first, second);
}

/*------------------------------------------------------------------
Module: swap()
Description: Exchanges the values stored in two variables.
Input Parms: The addresses of two integer variables.
Output Parms: The values of the input parameters are exchanged.
Returns: Nothing
Modifications
        Date              Description
        none              none
------------------------------------------------------------------*/
int cube(int number);
void swap(int *first_number, int *second_number)
{
    int temp_number;

    temp_number = *first_number;
    *first_number = *second_number;
    *second_number = temp_number;
}
```

Notice the use of the address-of operator in the call to swap(). Given the input values 3 and 5, the output of this program is

```
Enter two integers: 3 5
Your numbers swapped are 5 and 3
```

You have already been using a function that passes parameters by reference. Although you have not seen the definition for scanf(), the use of the address-of operator with its arguments tells you that its parameters are pointers.

Try It
♦ Define a function print_values() for the program in Example 4-12 that accepts the values of first and second from main() and prints them in the 24th and 48th columns of the current row on standard output. Design print_values() to pass-by-value.

♦ Define a function `get_values()` that prompts the user for two integers, reads the integers, and stores them in the `main()` variables `first` and `second`. Design `get_values()` to pass-by-reference.
♦ Modify `main()` so that, instead of calling `printf()` and `scanf()` from within the body of `main()`, it calls `get_values()` and `print_values()` in the appropriate place.

Application 1   FLUID VELOCITY

Civil Engineering

Civil engineers sometimes use the Manning formula to determine the velocity of fluid flowing out of the bottom of a reservoir through a pipe.

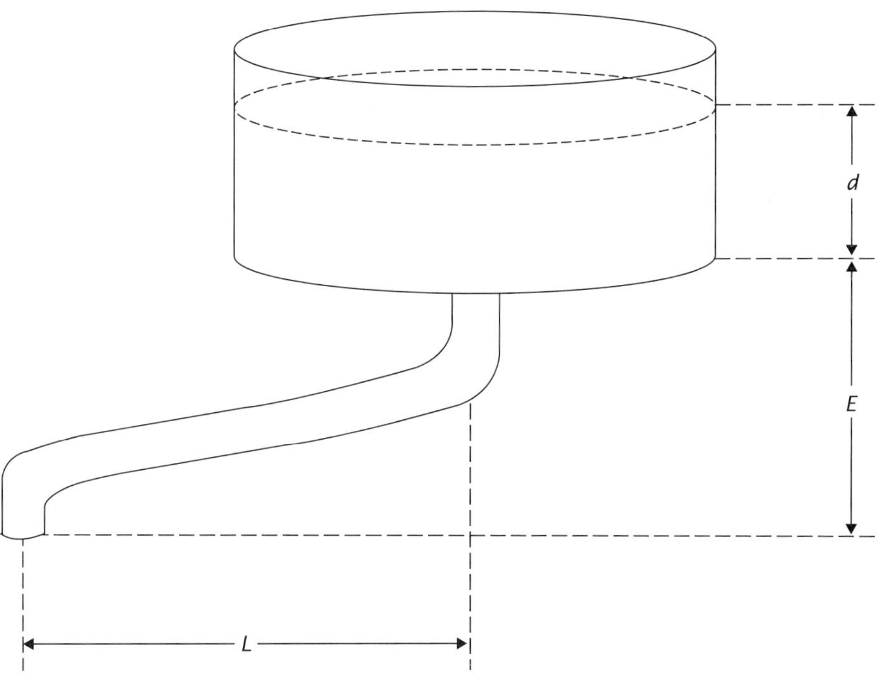

The Manning formula is

$$V = \frac{1.486}{F}R^{2/3}S^{1/2}$$

where

V = velocity in feet per second
F = friction coefficient

R = hydraulic radius of the pipe

(For this application R is simply the radius of the pipe, even though there are many applications where this is not the case.)

S = slope of the energy gradient = $\dfrac{d + E}{L}$ where E = elevation drop of the pipe, and L = length of pipe.

Develop a C program that will use this formula to help the user determine fluid velocity under similar circumstances.

1. Define the Problem

For this problem we are to write a C program that will prompt the user for the necessary information, accept inputs from standard input, apply the Manning formula to calculate fluid velocity, and print the fluid velocity on standard output.

2. Gather Information

The input data required by this program are

1. Friction coefficient — This is a floating-point value in the foot-pound-second system. As an example, the friction coefficient of cast iron is 0.015.
2. Hydraulic radius — For round pipe, the hydraulic radius is simply the pipe radius and is generally measured in inches.
3. Elevation drop of pipe — The vertical distance from the bottom of the reservoir to the end of the pipe is a real number and is generally measured in feet.
4. Length of pipe — This horizontal length of the pipe is a real number and is generally measured in feet.
5. Reservoir depth — The depth of the reservoir is generally measured in feet.

The output data is the fluid velocity as it exits the pipe. This value is a real number measured in feet per second.

3. Generate and Evaluate Potential Solutions

A potential solution to this problem can be diagrammed by the structure chart shown in Figure 4-6. A *structure chart* is a graphical representation of a problem solution. The problem-solving strategy is to break a complex problem into simpler parts. This strategy is known as *top-down design,* or *stepwise refinement.* The structure chart helps you determine the

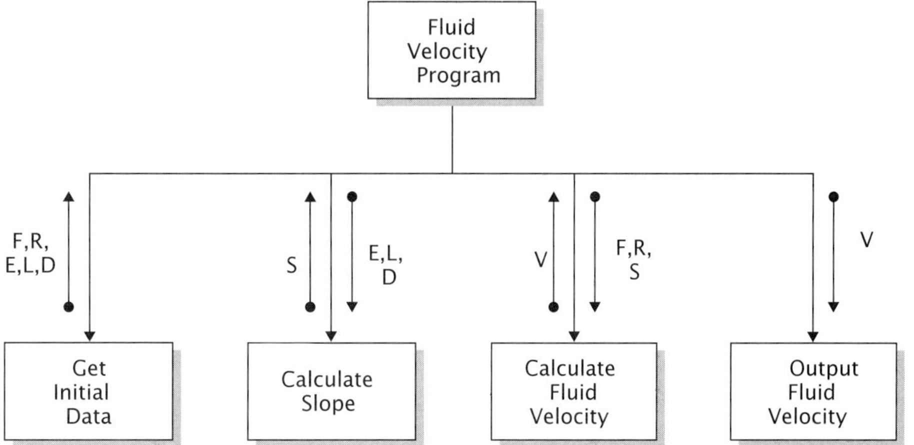

Figure 4-6 Fluid Velocity Program Structure Chart

modules that make up the program. Each of the boxes will potentially be implemented as a *C* function.

To properly calculate fluid velocity, the units of measure for *F*, *R*, and *S* must be the same. Because *V* is to be expressed in feet per second, it will be necessary to convert *R* to feet prior to calculating *V*.

5. Refine and Implement a Solution

The following program is a direct implementation of the design expressed by Figure 4-6. Look carefully at the statements that make up `main()`. Following the declaration of local variables, the next three statements are calls to functions. These functions correspond to the leftmost boxes in the structure chart. Because the rightmost box can be implemented by a single call to `printf()`, it is not necessary to put it into a function definition.

```
/*-------------------------------------------------------------
  Description: This program will calculate the velocity of a
      fluid flowing out of the bottom of a reservoir through
      a pipe using the Manning formula:

          V = (1.486/F)(R^(2/3))(S^(1/2)) where

          V = velocity in feet per second
          F = friction coefficient
          R = hydraulic radius of the pipe
          S = slope of energy gradient = (D+E)/L where
              E = elevation drop of pipe
              L = length of pipe

  Programmer: Ken Collier
  Date of Last Revision: 1-1-1995
```

```c
      Modifications
            Date              Description
            none              none
------------------------------------------------------------*/


#include <stdio.h>
#include <math.h>

/* Function Prototypes */
void get_initial_data(float *friction_coefficient,
                 float *hydraulic_radius,
                 float *elevation_drop,
                 float *pipe_length,
                 float *depth);

float calc_slope(     float elevation_drop,
                 float pipe_length,
                 float depth);
float calc_velocity  (float friction_coefficient,
                 float hydraulic_radius,
                 float slope);

void main(void)
{
   float    f_coeff,        /* Friction Coefficient */
            radius,         /* Hydraulic Radius */
            e_drop,         /* Elevation Drop of Pipe */
            length,         /* Length of Pipe */
            depth,          /* Depth of Reservoir */
            slope,          /* Slope of Energy Gradient */
            velocity;       /* Fluid Velocity */


   get_initial_data(&f_coeff, &radius, &e_drop, &length, &depth);
   slope = calc_slope(e_drop, length, depth);
   velocity = calc_velocity(f_coeff, radius, slope);
   printf("The fluid velocity in feet/sec is %f\n", velocity);
}


/*-----------------------------------------------------------
Module: get_initial_data
Description: This module prompts for and reads the input
     required for the calculation of fluid velocity.
Input Parms: The memory addresses of the uninitialized variables
     for friction coefficient, hydraulic radius, elevation
     drop, pipe length, and reservoir depth.
Output Parms: Each of the input parameters is initialized with
     meaningful information from standard input.
Returns: Nothing
Modifications
            Date              Description
            none              none
------------------------------------------------------------*/
```

```c
void get_initial_data(float *friction_coefficient,
                      float *hydraulic_radius,
                      float *elevation_drop,
                      float *pipe_length,
                      float *depth)
{
  printf("Enter friction coefficient: ");
  scanf("%f", friction_coefficient);
  printf("Enter hydraulic radius: ");
  scanf("%f", hydraulic_radius);
  printf("Enter elevation drop of pipe: ");
  scanf("%f", elevation_drop);
  printf("Enter pipe length: ");
  scanf("%f", pipe_length);
  printf("Enter reservoir depth: ");
  scanf("%f", depth);
}


/*-------------------------------------------------------------
Module: calc_slope
Description: This module calculates and returns the slope of
     energy gradient using the formula (D+E)/L
Input Parms: The values for D, E and L in the formula
Output Parms: None
Returns: Slope of Energy Gradient
Modifications
        Date             Description
        none             none
-------------------------------------------------------------*/
float calc_slope(    float elevation_drop,
                float pipe_length,
                float depth)
{
  return ((depth + elevation_drop)/pipe_length);
}


/*-------------------------------------------------------------
Module: calc_velocity
Description: This module calculates and returns the fluid
     velocity using the Manning formula.
Input Parms: The values for F, R and S in the formula
Output Parms: None
Returns: Fluid Velocity
Modifications
        Date             Description
        none             none
-------------------------------------------------------------*/
float calc_velocity (float friction_coefficient,
                float hydraulic_radius,
                float slope)
```

```
{
  float velocity, radius_in_feet;

  /* Convert Inches to Feet */
  radius_in_feet = hydraulic_radius/12.0;

  velocity = (1.486/friction_coefficient) *
             (pow(radius_in_feet, (2.0/3.0))) *
             (pow(slope, (1.0/2.0)));

  return(velocity);
}
```

Notice that the function `get_initial_data()` in this program passes parameters by reference, whereas the others pass by value. Because `get_initial_data()` returns five values according to the structure chart, passing by reference is necessary.

6. Verify and Test the Solution

We can test the program using a 5-foot-deep reservoir and a finished concrete pipe that is 12 inches in diameter (radius = 6 inches). The pipe is 20 feet long and drops 10 vertical feet from reservoir to exit. Finished concrete has a friction coefficient of 0.012. Given this input, the output of our program is

```
Enter friction coefficient: 0.015
Enter hydraulic radius: 6.0
Enter elevation drop of pipe: 10.0
Enter pipe length: 20.0
Enter reservoir depth: 5.0
The fluid velocity in feet/sec is 54.046993
```

Checking this input by hand tells us that the program works correctly for this data. Test this program against additional data to convince yourself that it works for a wide variety of input values.

What If

How could we redesign this program so that it works correctly but `get_initial_data()` does not pass parameters by reference? Would this change be an improvement over the existing program? Why, or why not?

SUMMARY

This chapter introduced you to modular programming with functions. Modular programming has long been considered paramount to good programming style by professional programmers; it greatly increases the ease of testing and debugging. Although you have been using predefined functions before now, in this chapter, you were introduced to user-defined functions, function prototypes, and function calls. Additionally, this chapter explored parameter passing by value and parameter passing by reference. The ability to define and use your own functions will increase your programming power. Good programmers take full advantage of the power provided by modular programming to create structured and well-organized programs. Programs that are unstructured can be very difficult to debug, maintain, and modify.

Key Words

actual argument	parameter
address-of-operator	parameter list
dereference	pass-by-reference
formal parameter	pass-by-value
forward reference	pointer
function call	program prologue
function prototype	return type specifier
global variable	scope
header file	side effect
local variable	stepwise refinement
module prologues	structure chart
modules	top-down design

Exercises

1. Another method of swapping the values of two variables uses a very elegant mathematical approach and avoids the need for a temporary variable. To swap the values stored in the variables x and y, simply apply the following algorithm:

   1. $y = x + y$
   2. $x = y - x$
   3. $y = y - x$

   Redefine the `swap()` function presented in this chapter to implement this algorithm. Does this change affect the program in Example 4-12?

2. In problem 9 of Chapter 3, you wrote a program to compute Reynold's number using the formula

$$R = \frac{\rho v D}{\mu}$$

   Now write a function with the prototype

   ```
   float reynolds_number (float rho, float v, float D, float mu);
   ```

   that will accept values for fluid density (ρ or rho), flow rate (v), pipe diameter (D), and fluid viscosity (μ or mu) and will return the result of the Reynold's number equation.

   Test your function by writing a `main()` function that will read the values for rho, v, D, and mu from standard input and will print the value returned from your function. Hand-check the results to make sure they are correct.

3. In problem 13 of Chapter 3, you wrote a program for converting Fahrenheit temperatures to Celsius. Define a function with the prototype

   ```
   float fahren_to_celsius(float temperature);
   ```

   that will accept a Fahrenheit temperature as its argument and will return its Celsius equivalent.

   Test your function by writing a `main()` function that will read a Fahrenheit temperature from standard input and will print its Celsius
   equivalent on standard output.

4. Modify your solution to problem 3 so that `fahren_to_celsius()` returns
   no value. Instead, define the function to change the value of its actual
   argument by using pass-by-reference parameters. Your new function
   prototype will look like

   ```
   void fahren_to_celsius(float *temperature);
   ```

   Test your function by modifying the `main()` function in problem 3 to
   accommodate the new function.

5. In problem 14 of Chapter 3, you wrote a program for converting Celsius
   temperatures to Kelvin. Define a function with the prototype

   ```
   float celsius_to_kelvin(float temperature);
   ```

   that will accept a Celsius temperature as its argument and will return its
   Kelvin equivalent.

   Test your function by writing a `main()` function that will read a Celsius temperature from standard input and will print its Kelvin equivalent on standard output.

6. Modify your solution to problem 5 so that `celsius_to_kelvin()` returns
   no value. Instead, define the function to change the value of its actual
   argument by using pass-by-reference parameters. Your new function
   prototype will look like

   ```
   void celsius_to_kelvin(float *temperature);
   ```

   Test your function by modifying the `main()` function in problem 5 to
   accommodate the new function.

7. Using the function definitions that you developed in problems 3 and 4,
   define a function with the prototype

   ```
   float fahren_to_kelvin(float temperature);
   ```

   that will accept a Fahrenheit temperature as its argument and will return its Kelvin equivalent. *Hint:* This function definition needs only to
   call other functions.

   Test your function by writing a `main()` function that will read a Fahrenheit temperature from standard input and will print its Celsius and
   Kelvin equivalents on standard output.

8. Modify your solution to problem 7 so that `fahren_to_kelvin()` returns no value. Instead, define the function to change the value of its actual argument by using pass-by-reference parameters. Your new function prototype will look like

```
void fahren_to_kelvin(float *temperature);
```

   Test your function by modifying the `main()` function in problem 5 to accommodate the new function.

9. Modify the program in the fluid velocity example to include a call to a function that calculates the Reynolds number.

10. The pressure in a liquid at any depth is determined by $P = P_0 + p \times g \times h$, where P_0 is the reference pressure, p is the fluid density, g is the acceleration due to gravity, and h is the depth below the surface. Write a program to calculate the pressure at a point on the dam at a depth from 0 to 100 feet by 10-foot increments. How could you develop this program to estimate the total force pressing on the dam face?

11. Ohm's law is crucial in calculating the behavior of electrical and electronic circuits, where Ohm's law is given by $V = I \times R$. Write a program that reads the circuit voltage and resistance and calculates the current. If the power is determined by $P = V \times I$, modify your program to calculate the resistance in a lightbulb with a given wattage (power) at standard U.S. voltage (120 VAC).

12. Resistors in series have an equivalent resistance of R = R1 + R2. Write a function to take two resistor values and return the equivalent resistance. Resistors in parallel have the equivalent resistance of 1/R = 1/R1 + 1/R2. Write a function to take two resistor values and return the equivalent parallel resistance. Write one function to take the values of two resistors and return both the equivalent resistance for parallel and series circuits. How could you use this in a program to calculate the equivalent resistance of three resistors in series; in parallel; two in series in parallel with the third; and two in parallel in series with the third? Could you use this program to determine the effective resistance of a large number of resistors?

13. Use the solutions to the unit-conversion problems in Chapter 3 to develop a library of programs to convert units using function calls.

5 Controlling the Flow Through Selection

Jumbo Jet The Boeing 777 is the newest in a series of commercial aircraft developed by Boeing in collaboration with design teams from many other companies worldwide. The 777 is controlled by several on-board computers, including flight-control, automatic-pilot, and instrument-monitoring systems. During a typical flight, the software in these computers must continually evaluate thousands of conditions and respond appropriately. To address this challenge developers designed the software using control structures and modularity. These techniques helped make the Boeing 777 the state-of-the-art aircraft that it is.

INTRODUCTION

So far your programs have consisted of sequences of statements executed in the order they appear. Many problems require that only certain statements be executed under certain circumstances, or that groups of statements be executed repeatedly. In this chapter you will learn how to control the flow of execution in your programs by specifying the conditions under which certain statements are executed. These constructs, known as *control structures,* form a fundamental set of tools required for writing powerful programs.

You will learn about the `if` statement, the `if-else` statement, and the `switch` statement. To use these statements effectively, you will be introduced to a new set of operators known as boolean operators and shown how you can use them to create logical expressions. Additionally, you will learn how to creatively combine control structures to write some elegant and powerful programs.

5-1 ADDING POWER TO YOUR ALGORITHMS

The program's flow of control refers to the natural order of execution of the statements that make up the program. Most typically, flow travels in a linear fashion from the top, or beginning of the program, to the bottom, or end, if it is not interrupted. One way to alter the default flow of control is through function calls. As you have seen, when a function is called in a program, the program's natural flow branches to a different segment of code known as the function definition. When the flow reaches the end of the function, it returns to the calling statement and continues onward.

Although functions provide a means of altering the flow, we require additional constructs. To solve all computable problems (some problems are not computable), you must control three aspects of the flow: sequencing, selection, and repetition. These concepts form the basis for the control structures provided by *C* and by most high-level languages.

Sequencing

Sequencing simply refers to your ability to dictate the order in which statements in a program will be executed; fortunately this is not difficult. In fact, you have been doing so already, perhaps without even realizing it. We typically think of a problem solution as a sequence of steps that arrive at some goal. Like a recipe, these steps must be performed in the proper order for the solution to succeed. The top-to-bottom default flow of control in a *C* program allows you to place program statements in the proper order. This is known as controlling the sequence and, because it is so intuitive, does not require any further explanation. Figure 5-1 is a flowchart that represents the idea of sequencing.

Selection

Under certain conditions, one set of statements may be appropriate to the problem, whereas under other conditions, a different set of statements is

Figure 5-1 A Flowchart Showing Sequencing

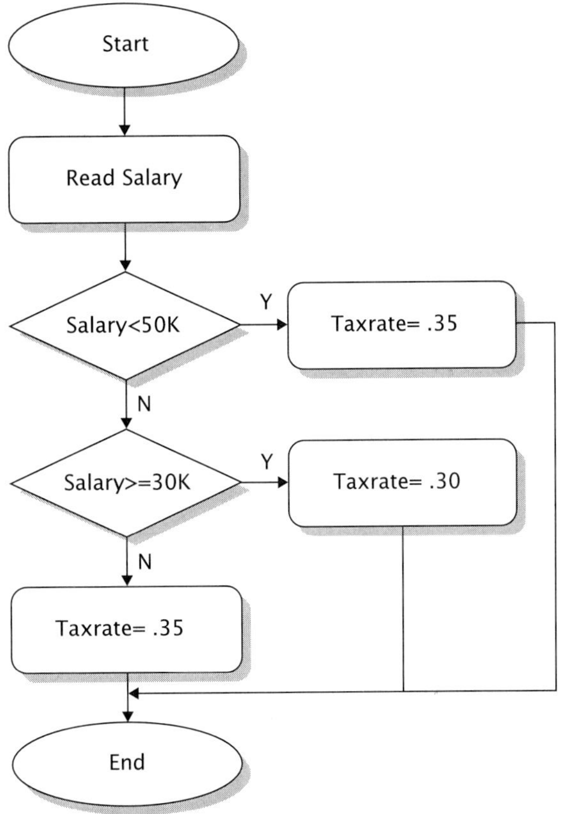

Figure 5-2 A Flowchart Showing Selection

appropriate. For example, consider a program to determine an employee's pay rate. If the salary is less than $30,000 per year, the employee is taxed at a 25 percent rate, whereas salaries between $30,000 and $50,000 are taxed at a 30 percent rate, and salaries above $50,000 are taxed at 35 percent. This situation requires that the program perform differently under different conditions. In Figure 5-2 we use a flowchart to show the conditional selection of tax rates. The ability to conditionally select a set of statements to be executed based on some conditions is called controlling the *selection*.

Repetition

It is often the case that the same statements must be executed multiple times. For example, the employee payroll program just described is useful only if it can be applied to every employee in the company. Although you could rewrite the same group of statements once for each employee, this is not a very elegant or attractive solution. It would be much better to specify that a grouping of statements execute repeatedly and that the number of repetitions be controlled. The ability to cause the flow of control to return to a previous point in the program and repeat a sequence of instructions is known as the control of *repetition*. In Figure 5-3 we use a flowchart to show repetition.

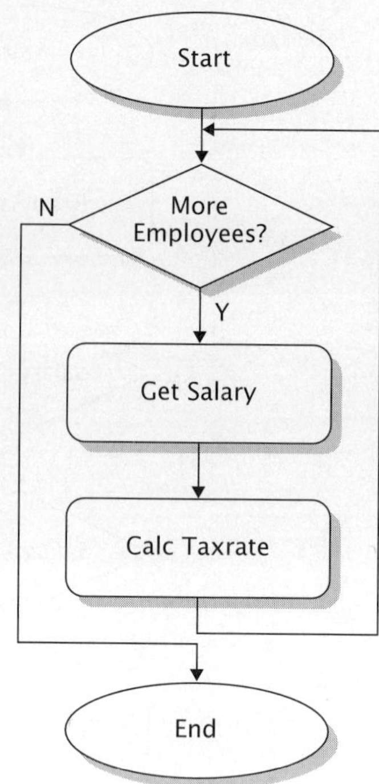

Figure 5-3 A Flowchart Showing Repetition

5-2 WRITING LOGICAL EXPRESSIONS

To describe the employee payroll tax rates, we used the following sentence:

"If the salary is *less than* $30,000 per year, the employee is taxed at a 25 percent rate, whereas salaries *between* $30,000 and $50,000 are taxed at a 30 percent rate, and salaries *above* $50,000 are taxed at 35 percent."

The italicized words in this sentence verbally describe mathematical relationships between salary amounts and tax-bracket boundaries. These relationships are represented in a program by boolean operators. A *boolean operator* is an operator that evaluates to either true or false, depending on the relationship between its operands. For example, the statement "$36,000 is less than $30,000" is false, whereas the statement "$36,000 is between $30,000 and $50,000" is true. The values *true* and *false* are known as *boolean values*.

Boolean operators fall into two categories, relational operators and logical operators. *Relational operators* are operators that describe relationships between their operands. *Logical operators* allow you to combine logical expressions into complex expressions. Relational operators have higher precedence than logical operators. Table 5-1 extends the table of operators (Table 3-1) by including the relational and logical operators in the precedence hierarchy. Boolean operators play a big role in the use of control structures, as you will soon see.

Table 5-1   Some of the Mathematical Operators in *C*

Category	Subcategory	Operator	Associativity
Parentheses		()	Innermost first, left-to-right
Arithmetic	Unary	-, ++, --, !	Right-to-left
	Binary first priority	*, /, %	Left-to-right
	Binary second priority	+, -	Left-to-right
Boolean	Relational (inequality)	<, >, <=, >=	Left-to-right
	Relational (equality)	==, !=	Left-to-right
	Logical	&&, ||	Left-to-right
Assignment		=, +=, -=, *=, /=, %=	Right-to-left

Relational Operators

Relational operators are further divided into the subcategories of *inequality operators* and *equality operators.* Inequality operators have a higher precedence than equality operators.

Inequality   The inequality operators include less than (<), less than or equal (<=), greater than (>), and greater than or equal (>=). These are all binary operators with left-to-right associativity.

We normally think about inequality expressions evaluating to true or false as described previously, but the *C* programming language does not recognize these boolean values. Instead, *C* uses the integer value 0 to represent false and a nonzero value (usually 1) to represent true. This implementation has some very interesting and useful ramifications, as we will see. Example 5-1 shows some examples of boolean expressions using inequality operators.

EXAMPLE 5-1

Boolean Expressions Using Inequality Operators

```
5 > 10                  /* False or 0 */
5 < 10                  /* True or nonzero value */
3.14159 >= 10           /* False */
'a' <= 'A'              /* False */
'A' <= 'a'              /* True: ASCII 65 <= ASCII 97 */
salary < 30000          /* Depends on the value of salary */
30000 > salary          /* Equivalent to: salary < 30000 */
(5 + 4) < (3 + 6)       /* False */
(10 * 5) >= (500 / 10)  /* True */
```

. . . .

Equality   Similar to the inequality operations, the equality operators equal (==) and not equal (!=) evaluate to zero or nonzero, depending on their truth value in an expression. Like the inequality operators, these are binary operators with left-to-right associativity. (*Warning*—A common programming error in *C* is the use of the assignment operator when the is-equal-to operator is intended. You will soon see why this is a problem.)

Table 5-2   Truth Table for the Relational Operators

X	Y	X<Y	X<=Y	X>Y	X>=Y	X==Y	X!=Y
10	10	0	1	0	1	1	0
14	23	1	1	0	0	0	1
31	29	0	0	1	1	0	1

EXAMPLE 5-2

Boolean Expressions Using Equality Operators

```
5 == 5                   /* True or nonzero value */
5 != 10                  /* True */
'a' == 97                /* True since ASCII value of 'a' is 97 */
'a' != 'A'               /* False since ASCII values are 97 and 65 */
salary == 30000          /* Depends on the value of salary */
(x + y) == 5             /* Depends on values of x & y */
(5 * 10) != (500/10)     /* False */
(100/10) != (50/10) /* True */
```

. . . .

Table 5-2 shows a truth table that further defines the behavior of the relational operators. A *truth table* shows all possible values for a mathematical or logical expression. Although the values assigned to X and Y are 0 and 1, the results in the truth table for different combinations of these values can be extrapolated to other numerical values with similar relationships to one another. Furthermore, the value 1 has been used in this truth table to represent the boolean value true for convenience sake. Although many *C* compilers use the value 1 to represent true, the ANSI standard specifies only that true be represented by any nonzero value.

Try It

Write a valid *C* statement to express each of the following statements:

- ◆ *x* plus *y* is greater than *y* plus *z*.
- ◆ *x* minus *y* is equal to *y* plus *z*.
- ◆ *y* is not strictly greater than *z*.
- ◆ *x* is not equal to *y* plus *z*.
- ◆ *x* is equal to the value of *y* after it has been incremented by 1.

Logical Operators

Suppose we modified the tax-rate conditions to account for number of dependents. For instance, if the salary is greater than $30,000 and the number of dependents is greater than 3, then the rate remains at 25 percent. To express this condition as a boolean equation, you need some means of connecting the two subexpressions. Logical operators serve this purpose. You can use logical operators to build complex boolean expressions by combining simpler expressions.

The logical operators are and (&&), or (||), and not (!). Logical operators are binary and have a left-to-right associativity. In the preceding condition, the connecting word is *and.* Therefore, you could write the condition in *C* as

```
(salary > 30000) && (dependents > 3)
```

In a complex condition built using &&, both operands must be true for the entire expression to be true. Conversely, an expression built using || is true if either operand is true. Consider the conditional statement, "Tax rate is 25 percent if salary is less than or equal to $30,000 *or* if the number of dependents is greater than 3." This condition can be written as the *C* expression

```
(salary <= 30000) || (dependents > 3)
```

The logical operator not (!) is a unary operator that precedes its operand; it has the same high precedence as other unary operators. It evaluates to true if its operand is false and false if its operand is true. This operator is used to express statements such as, "If you are not 21 or older, you cannot buy beer." This condition is stated in *C* as

```
!(age >= 21)
```

Example 5-3 shows some boolean expressions built using logical operators.

EXAMPLE 5-3

Boolean Expressions Using Logical Operators

```
!((5>6) || (6>5))               /* False */
(10 < (100/10)) && (5 < 9)      /* False */
(5<10) && ('a' == 97)           /* True */
!(5 < 6) && !(20 != 20)         /* True */
```

. . . .

The truth table shown in Table 5-3 further defines the behavior of the logical operators. Again, the value 1 to represent true is used simply for convenience. Understand that X and Y in this table do not necessarily represent simple variables containing integer values. They may represent the values of complex conditional subexpressions.

As with the arithmetic operators, you can combine relational and logical operators to form complex expressions whose result is either true (nonzero) or false (0). To express the constraint that salaries between $30,000 and $50,000 have a tax rate different from other salary ranges, you would

Table 5-3   Truth Table for the Logical Operators

X	Y	X&&Y	X||Y	!X&&Y
0	0	0	0	1
0	1	0	1	1
1	0	0	1	0
1	1	1	1	0

use the mathematical expression

$$30{,}000 \le \text{salary} \le 50{,}000$$

The same expression in *C* is

```
(salary >= 30000) && (salary <= 50000)
```

In *C*, this expression is actually two simple expressions combined with the && connective. Notice that we have surrounded the subexpressions with parentheses, which makes the expression easier to read and also ensures that the relational operations are evaluated before the logical operator. As with mathematical expressions, you should form the habit of using parentheses liberally.

Try It

Assuming that 1 represents the boolean value true and 0 the boolean value false, what is the value of the variable x after the execution of each of the following *C* expressions? Parenthesize these expressions so that they are more readable, but do not alter their meaning.

- x = 5 >= 3;
- x = 5 > 3 || 5 < 3;
- x = 5 > 3 && 5 < 3;
- x = −25 * 4 + 5 >= −4 + 5 * 25 && 50 + 50 < −2 * −50;
- x = −10 * −10 != 1000/10 || 25 * 25 + 50 == −50 * 25 + −25;

5-3   BRANCHING WITH THE SIMPLE if STATEMENT

So far you have seen conditional expressions by themselves. In a program, such expressions are typically used with a control structure to cause certain statements to execute if the condition is true. In this section you will learn to use conditional expressions in this way.

The simple if statement is the easiest of all the control structures to understand and use. The if statement is divided into a header and a body. The header contains the keyword *if* and a boolean expression. The body contains one or more lines of executable *C* code. If the boolean expression in the header is true, the code in the body will be executed. If it is false, the flow of control will branch to the statement following the code block. Figure 5-4 depicts the parts of an if statement.

The general form of the if statement is

```
if (<conditional expression>)
    <code block>
```

A classic example of the if statement involves writing a program that will read three integer values from the computer's keyboard into the variables low, middle, high and arrange them to ensure that low ≤ middle ≤ high. Because there is no guarantee the user will enter the values in the correct order, you must be prepared to exchange their values if necessary. This

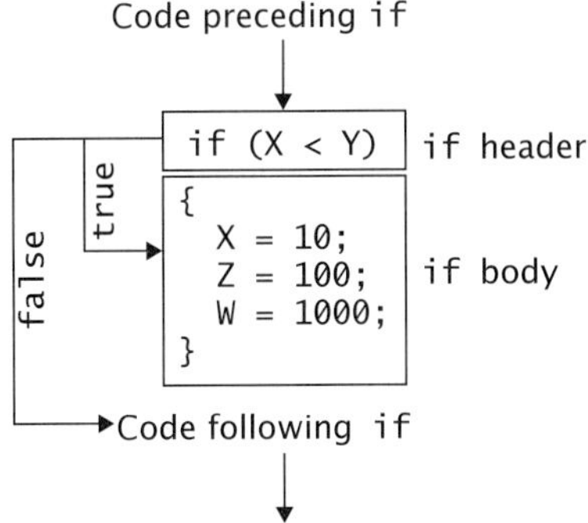

Figure 5-4 If Statement Structure

exchanging may be done in as few as three conceptual steps.

```
If low > middle then swap the contents of low and middle
If low > high then swap the contents of low and high
If middle > high then swap the contents of middle and high
```

The first two steps ensure that the smallest of the three values is stored in low. The final step in the algorithm simply sorts the two largest values into their proper positions.

The success of this algorithm hinges on the ability to swap the values of two variables. For this we can use the swap() function that was developed in Example 4-11. All that remains is to write a program to read three values, arrange them into ascending order, and print them out. Example 5-4 contains the complete program for doing this.

EXAMPLE 5-4   A Program to Arrange Three Numbers

```c
/*----------------------------------------------------------------
  Description: This program reads 3 numbers from standard
      input, arranges them into ascending order, and prints them
      on standard output.
  Programmer: Ken Collier
  Date of Last Revision: 1-1-1995
  Modifications
        Date            Description
        none            none
--------------------------------------------------------------*/
#include "stdio.h"

void swap(int *first_number, int *second_number);

void main(void)
{
  int low, middle, high;

  /* Read 3 numbers in any order */
  printf("Enter 3 integers: ");
  scanf("%d%d%d", &low, &middle, &high);
```

```
    /* Arrange them into ascending order */
    if (low > middle)
      swap(&low, &middle);         /* Sort first two values */
    if (low > high)
      swap(&low, &high);           /* Make sure low is smallest */
    if (middle > high)
      swap(&middle, &high);        /* Make sure high is largest */

    /* Print them in ascending order */
    printf("Your numbers in order: %d %d %d\n", low, middle, high);
}

/* Definition of swap() from Example 4-11 */
void swap(int *first_number, int *second_number)
{
    int temp_number;

    temp_number = *first_number;
    *first_number = *second_number;
    *second_number = temp_number;
}
```

. . . .

When given the input values 54, 31, and 99, this program prints

```
Enter 3 integers: 54 31 99
Your numbers in order: 31 54 99
```

The code block of the if statement can be simple or compound. A *simple code block* contains only a single instruction, as in the preceding examples. *Compound code blocks* contain two or more instructions and must be delimited by a pair of curly braces ({}) as with a function definition. Example 5-5 shows a program that checks to see if the user's salary is between $30,000 and $50,000. If it is, then the tax rate is set to 30 percent, and this rate is printed. Because the if body contains two statements, it is necessary to use curly braces. Note that we have placed the opening curly brace at the end of the if header. This placement helps keep the code from "spreading out" and becoming harder to read.

<table><tr><td>**EXAMPLE 5-5**</td></tr></table>

An if Statement with a Compound Body

```
#include <stdio.h>

void main(void)
{
    float salary,            /* To store user entered salary */
          taxrate=.25;       /* Default taxrate is 25% */

    /* Read salary amount */
    printf("Enter salary: ");
    scanf("%f", &salary);
```

```
  /* Update taxrate if necessary */
  if ((salary >= 30000) && (salary <= 50000)){
    taxrate = .30;
    printf("Taxrate is 30%%\n");
  } /* End if */

  /* Print net pay */
  printf("Take home will be %f.\n", salary*(1.0-taxrate));
}
```

. . . .

Given inputs of 35,000 and 25,000 on two consecutive executions, the output of this program is

```
Enter salary: 35000
Taxrate is 30
Take home will be 24500.00.
Enter salary: 25000
Take home will be 18750.00.
```

There are very few limitations on what can occupy the code block of an if statement. In fact, virtually any segment of code that is valid elsewhere in a program may reside in the block of an if statement.

You also can use an if statement to detect errors during a program's execution. For example, consider a program that reads the age of the user from standard input using the statements

```
printf("Enter your age: ");
scanf("%d", &age);
```

It is possible for the user to enter an invalid value such as a negative number or an unrealistically large number. You can use the following if statement to check for these errors and print an error message if an error is detected:

```
if ((age < 0) || (age > 120))
   printf("Error - value invalid.\n");
```

In this example, age is the control variable, because its value determines the truth of the conditional expression. If the condition is true, this statement prints an error message to the computer's display. This error-detection scheme does nothing to correct or handle the problem, though. One possibility is for you to abort the program when an error occurs. To do this will require a compound block in the if statement.

```
if ((age < 0) || (age > 120)){
   printf("Error - value invalid.\n");
   exit(1);
} /* End If */
```

The exit statement is similar to a return statement except that it exits from the entire program. Generally, if a program halts normally (with no errors), the exit value is 0. If the program halts abnormally (because of an error), the exit value is 1.

This error-handling approach is quite harsh. It might be more reasonable to give the user a chance to reenter a correct value. In Chapter 6 you will learn how to use repetition to detect and handle input errors.

Example 5-6 shows a program that reads the user's age and the current year, checks for errors in the user's age, and prints the user's birth year.

EXAMPLE 5-6   Using `if` Statements to Detect Errors

```
/*--------------------------------------------------------------------
   Description: This program prints the year the user was born
       given the current year and the user's age as inputs.
   Programmer: Ken Collier
   Date of Last Revision: 1-1-1995
   Modifications
         Date              Description
         none              none
-------------------------------------------------------------------*/
#include <stdio.h>

void main(void)
{
  int year, age;

  /* Read age */
  printf("Enter your age followed by the current year: ");
  scanf("%d%d", &age, &year);

  /* Check for error */
  if ((age < 0) || (age > 120)){
    printf("Error - invalid age - reenter: ");
    scanf("%d", &age);
  } /* End If */

  /* Print birth year */
  printf("You were born in %d.\n", year-age);
}
```

. . . .

Given an age of 54 and the year 1995, the output of this program is

```
Enter your age followed by the current year:
54 1995
You were born in 1941.
```

Try It   Modify the program in Example 5-6 to detect errors in the year as well as the age. The program should not accept negative-numbered years or years with five digits.

Avoid Mixing Assignment and Equality Operators

Earlier we warned you about using the assignment operator in place of the equality operator. An example of this mistake is shown in the following `if` statement:

```
if (x = 5)
    y = x++;
```

Although this statement appears to be correct, there is something wrong with the code. The assignment operator evaluates to the value expressed by its right-hand-side operand. In this example, the expression x = 5 assigns the value 5 to x and evaluates to 5. Because 5 is a nonzero value, the `if` condition is considered to be true. As a result, y is assigned the value of x (5), and x is incremented by 1. These events take place regardless of the initial value of x. But the original intent of this `if` statement is that y and x should change if x is equal to 5. Clearly the actual behavior is different from the intended behavior. This mistake is due to the use of = instead of ==. Be careful to avoid making this common mistake.

Try It   Many theaters give a 10 percent discount to people who are 55 years of age or older. Write a simple `if` statement that will set the value of a float variable called `discount` to 0.10 if the value of the integer variable age is 55 or greater.

5-4 BICONDITIONAL BRANCHING WITH `if-else`

Handling the tax-rate example using only simple `if` statements requires a separate structure for each condition, such as

```
if (salary < 30000)
    taxrate = 0.25;
if (salary >= 30000) && (salary <= 50000))
    taxrate = 0.30;
if (salary > 50000)
    taxrate = 0.35;
```

This approach is inefficient because all three conditions will be tested even though only one at a time will ever be true. The `if-else` construct allows you to combine these statements into a more efficient control structure. The `if-else` statement is also known as a *biconditional* because it has two code blocks, one of which will always be executed. The general form of the `if-else` statement is as follows:

```
if (<condition>)
    <if code block>
else
    <else code block>
```

The first two conditions in the previous example can be written using a single `if-else` statement.

```
(salary < 30000)
    taxrate = 0.25;
else
    taxrate = 0.30;
```

If the condition is true, the `if` body is executed and the `else` body is ignored. If the condition is false, the `if` body is ignored and the `else` body is executed.

Suppose you wish to write a program to find the largest number of three numbers entered by the user. There is no guarantee that the user will enter the numbers in any particular order, so you must compare the numbers to find the largest. The algorithm for this program is relatively simple.

```
Read three numbers.
Compare the first two numbers and keep the larger.
Compare the number kept in step one with the last number and keep
the larger.
Print the largest.
```

Example 5-7 contains a program that uses the `if-else` structure to implement this algorithm.

EXAMPLE 5-7   ## Finding the Largest of Three Numbers

```c
/*---------------------------------------------------------------
   Description: This program will find the largest of three
      numbers entered by the user.
   Programmer: Ken Collier
   Date of Last Revision: 1-1-1995
   Modifications
        Date                Description
        none                none
-----------------------------------------------------------------*/
#include <stdio.h>

void main(void)
{
  int num1, num2, num3,      /* Numbers input by user */
      largest;               /* Stores largest number */

  /* Read numbers */
  printf("Enter 3 integers: ");
  scanf("%d%d%d", &num1, &num2, &num3);

  /* Compare first 2 and keep the larger */
  if (num1 > num2)
    largest = num1;
  else
    largest = num2;
```

```
/* Now compare the larger with the third, keep largest */
if (num3 > largest)
   largest = num3;

printf("Largest is %d.\n", largest);
}
```

. . . .

Given input values 33, 53, and 233, the ouput of this program is

```
Enter 3 integers: 33 53 233
Largest is 233.
```

The first step in the algorithm has two possible actions, depending on the values of num1 and num2. Copy either num1 or num2 into largest. The if-else structure is well suited to solving this step.

In this example, the if body and the else body are simple statements and do not require curly braces surrounding them. However, as with the simple if statement, you can use curly braces for compound statements.

The if and if-else structures can be combined to perform multiway branching when there are more than two alternative sets of actions to perform. An entire if or if-else structure is viewed by the compiler as a single C statement. For this reason, either can reside anywhere within a program that a C statement can reside. Therefore, an if or if-else construct can reside within the body of another if or if-else construct.

As an example, consider a program to determine how many years until the next leap year. Simplistically stated, a leap year is a year that is divisible by 4. Suppose you wish to write a program that asks the user for the year and reports when the next leap year will occur. The modulus operator helps determine if the year entered is divisible by 4. Example 5-8 contains the program to calculate the number of years until the next leap year.

EXAMPLE 5-8

Calculating the Number of Years Until the Next Leap Year

```
/*------------------------------------------------------------------
   Description: This program will print the number of years until
      the next leap year given a year entered by the user.
   Programmer: Ken Collier
   Date of Last Revision: 1-1-1995
   Modifications
         Date                Description
         none                none
   ------------------------------------------------------------------*/
#include <stdio.h>

void main(void)
{
   int year;   /* Entered by user */
```

```
    /* Read year */
    printf("Enter a year: ");
    scanf("%d", &year);

    /* Print number of year until the next leap year */
    if (year % 4 == 0)
      printf("This is a leap year.\n");
    else if (year % 4 == 1)
      printf("Three years until the next leap year.\n");
    else if (year % 4 == 2)
      printf("Two years until the next leap year.\n");
    else
      printf("One year until the next leap year.\n");
}
```

. . . .

Given the input value 1995, the output of this program is

```
        Enter a year: 1995
        One year until the next leap year.
```

The compound `if` statement in this example is called a *cascading* `if` *structure* because control cascades through the structure until a condition is true or until the final `else` is reached. Figure 5-5 provides a means of visualizing how these `if-else` statements are nested. The block labeled B forms the entire body of the `else` clause of the `if-else` structure in the block labeled A. Likewise, the block labeled C forms the body of the `else` clause of B's `if-else` structure.

When the control flow reaches the first `if` condition in the sample execution, the condition is false and control branches to the `else` part of clause A to execute its code block. Because this block is itself an `if` statement, its condition is evaluated. Because this condition is also false, control branches to B's `else` clause and executes its code block. Again, this block is an `if-else` statement, and its condition is evaluated. This condition is also false so control branches to C's `else` clause and executes its code block, which is a simple call to `printf()`.

Note that in Example 5-8 we placed each new `if` statement on the same line as the `else` clause for which it is a code block. Furthermore, we did not indent each new `if-else` structure as in Figure 5-5. *C* does not care about

```
if (year % 4 == 0)
  printf ("This is a leap year.\n");                              A
else
    if (year % 4 == 1)
      printf("Three years until the next leap year.\n");        B
    else
        if (year % 4 == 2)
          printf("Two years until the next leap year.\n";     C
        else
          printf("One year until the next leap year.\n");
```

Figure 5-5 Visualizing a Cascading `if` Structure

indentation or whether each new statement resides on a new line; either style is acceptable. However, for large cascading `if` structures, indenting each new structure tends to march your statements off the right edge of the screen. Including each new `if` on the same line as the previous `else` helps vertically shrink the code, allowing more code to appear on the screen. For both of these reasons, the style used in Example 5-8 enhances readability. Whatever style you choose for your code, be sure that it makes the code as easy to read as possible.

You can use a cascading `if` structure to write the complete tax-rate calculation.

```
if (salary < 30000)
   taxrate = 0.25;
else if (salary <= 50000)
   taxrate = 0.30;
else
   taxrate = 0.35;
```

Notice that it is not necessary to express the second condition as

```
(salary >= 30000) && (salary <= 50000)
```

because the only way control will arrive at this condition is if the first condition is false.

In general, `if` and `if-else` statements can be nested together in an unlimited variety of ways. The key rule when nesting these constructs is that each new construct must be completely contained within the code block of an `if` or `else` clause.

Try It   Hand-trace the program in Example 5-8 using the dates 1997, 1998, and 1999. Which `if` conditions are evaluated, and which are ignored?

What If   An engineering firm pays its employees a 5 percent bonus for the first 9 years of employment and a 10 percent bonus after they have been with the firm for ten years. Write an `if-else` expression that will set the variable bonus equal to 0.05 if the value of the variable employment is between 0 and 9, and 0.10 if it is 10 or more. Try writing equivalent code using only a simple `if` statement.

5-5   MULTIWAY BRANCHING WITH THE switch STATEMENT

Very large cascading `if` statements can become difficult to read and understand. For this reason, *C* supports another control structure called the `switch` statement. As you saw, the `if` and `if-else` statements are very useful when it is possible to write one or two conditions to describe a wide variety of possibilities. The `switch` statement is a multiway branching statement that is useful when the control variable is being tested against a discrete set of possible values. The general form of the `switch` statement is

```
switch(<controlling expression>){
    case <constant1>:
        [<code block>;]
        [break;]
    case <constant2>:
        [<code block>;]
        [break;]
    case <constant3>:
        [<code block>;]
        [break;]
            .
            .
            .
    [default:
        <code block>;]
}
```

The expression in the `switch` statement is evaluated and tested for equality with the constant in each of the `case` clauses. If the value of the expression is equal to the constant in a `case` clause, then the code block immediately following the `case` clause will be executed. It is important to understand that the only test being made in a `switch` statement is a test for equality.

The `switch` statement provides a different way to solve the leap-year problem you encountered in Example 5-8. Example 5-9 demonstrates this new approach.

EXAMPLE 5-9   Reporting Leap Years Using `switch`

```
/*-------------------------------------------------------------------
   Description: This program will print the number of years until
      the next leap year given a year entered by the user.
   Programmer: Ken Collier
   Date of Last Revision: 1-1-1995
   Modifications
         Date              Description
         none              none
-----------------------------------------------------------------*/
#include <stdio.h>

void main(void)
{
  int year;

  printf("Enter the year: ");
  scanf("%d", &year);
```

```
switch(year % 4){
  case 0:
    printf("This is a leap year\n");
    break;
  case 1:
    printf("Three years until the next leap year.\n");
    break;
  case 2:
    printf("Two years until the next leap year.\n");
    break;
  default:
    printf("One year until the next leap year.\n");
    break;
}
}
```

· ·

Again, the output of this program given the input year 1995 is

```
Enter the year: 1995
One year until the next leap year.
```

Like the cascading `if` structure, when control reaches the `switch` statement, the expression is evaluated. Control then flows through the `switch` body and examines each `case` clause in search of a value that matches the value of the expression. When a match is found, the body of the matching `case` clause is executed. Unlike the `if` statement, after the body of a `case` clause is executed, control continues to flow through the remainder of the `switch` body in search of other matching values. This continued search is generally a waste of processor time and is inefficient. For this reason, the optional `break` statement should be used in all `case` clauses. The `break` statement causes the flow of control to branch to the `}` at the end of the construct.

Four rules apply to the `switch` statement.

1. The expression must evaluate to one of the integral *C* types (`char` or `int`).
2. The expression can contain a variable, a constant, or an expression.
3. The constant that is contained in each `case` clause must be either a literal constant (for example, 5 or a), a symbolic constant (for example, `MY_CONSTANT`), or an expression containing only constants (for example, `5 + MY_CONSTANT`).
4. No two `case` constant expressions may evaluate to the same value in the same `switch` statement.

The optional `default` clause at the end of the `switch` statement always evaluates to true, so its code block will always execute. This is another reason for the use of the `break` statement within each `case` clause. Without the `break` statement, a match might be made with one of the `case` clauses, and then another match would be made with the default clause.

You can combine `if` and `if-else` statements with `switch` statements. An `if` or `if-else` statement can reside in the body of a `case` clause, or an entire `switch` statement can reside in the body of an `if` or `if-else` statement. Furthermore, you can place a `switch` statement within the `case` clause of another `switch` statement.

Try It    Write a `switch` statement that will print a comment to the computer's display based on the user's letter grade as follows:

♦ If the grade is an A print "Excellent work. You earned an A."
♦ If the grade is a B print "Nice job. You earned a B."
♦ If the grade is a C print "You passed with a C."
♦ If the grade is a D print "You didn't pass, but you earned a D."
♦ Otherwise, print "Try again."

Application 1   IDENTIFYING MILITARY AIRCRAFT

Aerospace Engineering

Military aircraft are divided into classes of fighters, bombers, cargo planes, trainers, and so on. Models within each class are assigned identification numbers. Each aircraft has a set of distinguishing characteristics described in Table 5-4.

Develop a *C* program to help the user recognize the identifying characteristics of each aircraft model.

---

1. Define the Problem

You are to write a program that will ask the user for the class and ID of an aircraft model and that will print the distinguishing characteristics of that model.

Table 5-4   Identifying Military Aircraft

Class	ID	Characteristics
F	15	Twin Tail, Twin Engine
F	16	Single Tail, Single Engine
B	1	Sweeping Wings, Four Engines
B	2	Flying Wing
B	52	Eight Engines, Smokes on Take-off
C	5	Huge, Four engines
C	130	Four Engines, Propeller, High Wing
C	141	Four Engines, Propeller, High Wing
T	37	Two passenger, Single Engine, Blunt Nose
T	38	Supersonic, Needle Nose, Two Passenger
T	41	Propeller, High Wing, Looks like Cessna 150

 2. Gather Information

The input data required by this program are

1. Class label—This is a single character from the set {F, B, C, T}. We must design the program to accept both lowercase and uppercase characters. The program should also check for input errors.
2. Aircraft ID—This data item is an integer from the set {1, 2, 5, 15, 16, 37, 38, 41, 52, 130, 141} representing the aircraft model number.

The output of this program is a set of identifying characteristics displayed on standard output.

3. Generate and Evaluate Potential Solutions

A potential solution to this problem can be expressed by the structure chart shown in Figure 5-6.

The `get_aircraft_class()` function will be responsible for reading the class from standard input and returning a valid class value. This function must be able to accept both uppercase and lowercase characters and must detect invalid characters. Although this function could return the class value in whichever case it was entered, it will simplify other parts of the program if `get_aircraft_class()` always returns uppercase characters. We must decide how to handle this case conversion. The *C* library ctype (for character types) supports a collection of functions to manipulate `char` data. Two of these functions are named `toupper()` and `tolower()`. The first of these accepts a single character argument and returns its uppercase equivalent. If it is already in uppercase or if it does not have an uppercase equivalent, `toupper()` simply returns the character unchanged. The function `tolower()` works similarly for lowercase conversion. You can use either of these functions to ensure that the case

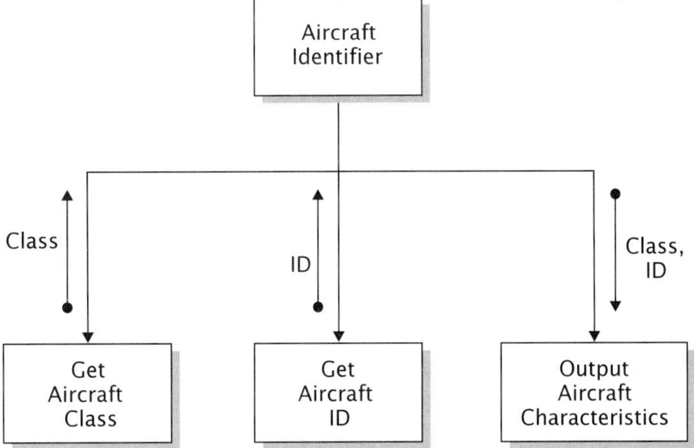

Figure 5-6 Aircraft Identifier Structure Chart

of the character returned from the `get_aircraft_class()` module is consistent. You will also handle error checking within the modules using an `if` statement.

The `get_aircraft_id()` function works in much the same manner as `get_aircraft_class()`. It must read the ID number from standard input and check for errors. For error handling you will avoid writing the complicated conditional expression required to test for each valid ID number. Instead, you will ensure the ID number is positive and allow the `print_aircraft_characteristics()` module to handle invalid ID numbers.

The `print_aircraft_characteristics()` will use a set of nested `switch` statements to determine which characteristics to print. The first-level `switch` statement will isolate the aircraft class, and the second-level `switch` statements will handle the ID numbers within the class.

4. Refine and Implement a Solution

The following program is a direct implementation of the design expressed by Figure 5-6. Note that the primary workhorse in this program is the function `print_aircraft_characteristics()`. It contains a large `switch` statement for handling the aircraft class. Within each `case` clause of this `switch` statement is either a cascading `if` structure or a smaller `switch` statement for handling the aircraft ID number. Study this function carefully to be certain you understand how these control structures are working together.

```c
/*---------------------------------------------------------------
  Description: This program will print the distinguishing
     characteristics of several military aircraft.
  Programmer: Ken Collier
  Date of Last Revision: 1-1-1995
  Modifications
        Date              Description
        none              none
----------------------------------------------------------------*/
#include <stdio.h>
#include <ctype.h>

char get_aircraft_class(void);
int get_aircraft_id(void);
void print_aircraft_characteristics(char class, int id);

void main(void)
{
  char class;/* Aircraft class from set {F,B,C,T} */
  int id; /* Aircraft ID */

  class = get_aircraft_class();
  id = get_aircraft_id();
  print_aircraft_characteristics(class, id);
}
```

```c
/*-------------------------------------------------------------
Module: get_aircraft_class()
Description: Reads aircraft class from standard input, checks
  for errors, converts class to uppercase, and returns class.
Input Parms: none
Returns: Character representing an aircraft class.
Modifications
        Date            Description
        none            none
-------------------------------------------------------------*/
char get_aircraft_class(void)
{
  char response;

  /* Read response */
  printf("Enter the class of the aircraft {F, B, C, or T}: ");
  scanf("%c", &response);

  /* Convert to uppercase */
  response = toupper(response);

  /* Check for errors - Assume second response is correct */
  if ((response != 'F') &&
      (response != 'B') &&
      (response != 'C') &&
      (response != 'T')){
    printf("Invalid class, reenter: ");
    scanf("%c", &response);
    response = toupper(response);
  } /* end if */

  return(response);
}


/*-------------------------------------------------------------
Module: get_aircraft_id()
Description: Reads aircraft ID from standard input, checks
  for errors, and returns ID.
Input Parms: none
Returns: Integer representing an aircraft ID.
Modifications
        Date            Description
        none            none
-------------------------------------------------------------*/
int get_aircraft_id(void)
{
  int id;

  /* Read aircraft id */
  printf("Enter aircraft ID number: ");
  scanf("%d", &id);
```

```c
    /* Error check to ensure positive number */
    if (id <= 0){
      printf("Invalid ID number, reenter: ");
      scanf("%d", &id);
    }

    return(id);
}


/*------------------------------------------------------------------
Module: print_aircraft_characteristics()
Description: Prints the distinguising characteristics of an
  aircraft model.
Input Parms: Class and ID number of target aircraft.
Returns: none
Modifications
        Date            Description
        none            none
------------------------------------------------------------------*/
void print_aircraft_characteristics(char class, int id)
{
  /* First use switch to handle class */
  switch (class){
    case 'F':
      /* Then use if to handle id - could use switch again */
      if (id == 15)
      printf("F-15: Twin Tail, Twin Engine\n");
      else if (id == 16)
      printf("F-16: Single Tail, Single Engine\n");
      else
      printf("Unknown aircraft\n");
      break;
    case 'B':
      /* This time use switch to handle id */
      switch (id){
      case 1:
        printf("B-1: Sweeping Wings, Four Engines\n");
          break;
      case 2:
        printf("B-2: Flying Wing\n");
          break;
      case 52:
        printf("B-52: Eight Engines, Smokes on Take-off\n");
          break;
      default:
        printf("Unknown aircraft\n");
        break;
      } /* end switch */
      break;
    case 'C':
      /* Back to the cascading if */
      if (id == 5)
      printf("C-15: Huge, Four engines\n");
      else if (id == 130)
```

```c
      printf("C-130: Four Engines, Propeller, High Wing\n");
       else if (id == 141)
      printf("C-141: Four Engines, Looks like Small C-5\n");
       else
      printf("Unknown aircraft\n");
       break;
    case 'T':
       if (id == 37)
      printf("T-37: Two passenger, Single Engine, Blunt Nose\n");
       else if (id == 38)
      printf("T-38: Supersonic, Needle Nose, Two Passenger\n");
       else if (id == 41)
      printf("T-41: Propeller, High Wing, Like Cessna 150\n");
       else
      printf("Unknown aircraft\n");
       break;
    default:
       printf("Invalid class\n");
       break;
  }
}
```

5. Verify and Test the Solution

This program is not difficult to test exhaustively. You should input each aircraft class and ID number to see if the results are what you expected. Additionally, you should enter some uppercase and some lowercase characters for the aircraft class. To be complete, you should also test the program with combinations of invalid classes with valid IDs, valid classes with invalid IDs, and invalid classes with invalid IDs. Given the data combinations p-38, F-15, C-141, t-37, b-46, and F-18, the output of this program is

```
Enter the class of the aircraft {F, B, C, or T}: p
Invalid class, reenter: p

Enter aircraft ID number: 38
Invalid class

Enter the class of the aircraft {F, B, C, or T}: F
Enter aircraft ID number: 15
F-15: Twin Tail, Twin Engine

Enter the class of the aircraft {F, B, C, or T}: C
Enter aircraft ID number: 141
C-141: Four Engines, Looks like Small C-5

Enter the class of the aircraft {F, B, C, or T}: t
Enter aircraft ID number: 37
T-37: Two passenger, Single Engine, Blunt Nose
```

```
Enter the class of the aircraft {F, B, C, or T}: b
Enter aircraft ID number: 46
Unknown aircraft

Enter the class of the aircraft {F, B, C, or T}: f
Enter aircraft ID number: 18
Unknown aircraft
```

Try It   Hand-trace an execution of this program with the first input in the test set. Why does the program allow you to enter an ID number when P is an invalid class?

Summary   This chapter introduced you to *C*'s relational and logical operators. These operators provide the necessary tools for writing expressions to test the truth of certain conditions. You also learned that sequencing, selection, and repetition are necessary to solve all computable problems. You explored the details of the selection-control structures provided by *C*. The `if`, `if-else`, and `switch` statements provide a variety of selection constructs. These control structures can be combined to form very powerful programs.

The contents of this chapter mark a turning point in your ability to solve more complex problems. The understanding of selection-control structures allows you to solve more difficult problems in very elegant ways.

Key Words

biconditional	logical operators
boolean operators	relational operators
boolean values	repetition
cascading `if` structure	selection
compound code block	sequencing
control structures	simple code block
equality operators	truth table
inequality operators	

Exercises

1. Write a program that reads a number from standard input and uses three simple `if` statements to print whether the number is negative, 0, or positive on standard output.

2. Modify your solution to problem 1 to use a cascading `if` structure rather than simple `if` statements.

3. Earlier the test for a leap year was oversimplified. The divisibility-by-4 test is only valid for noncentennial years. A centennial year (a year divisible by 100) must also be divisible by 400 in order to be a leap year. For that matter, millennial years (divisible by 1000) must also be divisible by 4000 in order to be leap years. Write a *C* program that reads a year as

a four-digit integer from standard input and uses an `if-else` statement to print whether the year is a leap year.

4. In civilian time notation, times of day are succeeded by midnight, a.m., noon, or p.m. Military time dispenses with these categorizations by ranging from 0000 (midnight) to 2359 (11:59 p.m.). Write a *C* program that reads the time in military format from standard input as a four-digit integer and uses a cascading `if-else` statement to determine whether the time is a midnight, a.m., noon, or p.m. time.

5. Sounds are produced by mechanical vibrations. These vibrations excite the air particles around the source and sound travels and can then be heard by other listeners. A sound's frequency is the number of times its vibration is repeated per second. This frequency is measured in Hertz (Hz) units. 1Hz = 1 cycle per second. The human ear is capable of detecting frequencies between about 20 and 20,000Hz. Below 20Hz is the infrasonic range, and above 20,000Hz is the ultrasonic range. We can categorize sounds into very low (20-5,000Hz), low (5,000-10,000Hz), medium (10,000-15,000Hz), and high (15,000-20,000Hz) pitches. Write a program that will read a sound frequency from standard input and will use a cascading `if` statement to identify it as low, medium, or high frequency. Print the result on standard output.

6. Sound amplitude, or loudness, is measured in decibel (dB) units. In this decibel scale, a 10-decibel (dB) difference between two sounds is perceived as a loudness difference of a factor of two. For example, 20dB is twice as loud as 10dB. Continuous or regular exposure to sounds of 90dB will eventually cause hearing loss, and sounds of 130dB or more can cause immediate hearing loss. We can categorize loudness as follows: ambient noise (0-30dB), invasive noise (30-70dB), irritating noise (70-90dB), aggravating noise (90-110dB), deafening noise (110dB and above). Develop a *C* program that reads sound amplitude in dB from standard input and uses a cascading `if` statement to print the sound classification to standard output.

7. Combine your solutions to problems 5 and 6 to develop a program that will read both frequency and amplitude and that will characterize the sound in qualitative terms.

8. In Chapter 4 you wrote several functions to convert temperatures between Fahrenheit, Celsius, and Kelvin scales. Write a *C* program that reads a temperature, an initial scale, and a target scale from standard input; converts the temperature from the initial scale to the target scale; and displays the converted temperature on standard output. Scales should be input as single characters (k = Kelvin, c = Celsius, f = Fahrenheit). Your program should accept both uppercase and lowercase characters. *Hint:* Use a `switch` statement with nested `switch` statements to solve your problem.

9. Suppose that water costs $0.55 per gallon and that a surcharge of 10 percent is added to the bill. The city also assesses a utility tax of 4 percent; however, the surcharge is not taxed, and the tax is not factored into the surcharge. Develop a program that will read the number of gallons used during a given month and that will calculate the amount to be billed for that month.

10. Civil engineers who are concerned with moving a fluid such as oil through a pipe from one point to another are interested in calculating the drop in pressure as a result of the friction within the pipe. Assuming the fluid is pumped at a steady rate, the pressure drop is given by $dP = P_A - P_B = \rho(gh + W)$, where

ρ = fluid density

W = energy lost per kilogram due to the fluid friction and is described by $W = \dfrac{4fv^2L}{D}$

g = gravitational acceleration

h = elevation change from P_A to P_B

L = length of pipe in meters

D = diameter of pipe in meters

v = flow rate in meters per second

f = friction factor. If the pipe is smooth, then the friction factor depends only on the Reynolds's number R, which is given by $R = \dfrac{\rho v D}{\mu}$, where μ is fluid viscosity.

If $(R < 2000)$, the flow is laminar and $f = 8/R$.

If $(R \geq 2000)$, the flow is turbulent and

$$f = .0395R^{-1/4}$$

In problem 2 of Chapter 4 you wrote a function to calculate Reynold's number. Write a function for the prototype

```
float friction_factor(float rho, float v, float D, float mu);
```

that will accept the fluid density (rho), flow rate (v), pipe diameter (D), and fluid viscosity (mu) as input arguments and will return the friction factor based on the following conditions:

If $(R < 2000)$, the flow is laminar and $f = 8/R$.

If $(R \geq 2000)$, the flow is turbulent and

$$f = .0395R^{-1/4}$$

Design the function to call the function you wrote to solve problem 2 in Chapter 4.

11. Using your solution to problem 10, write a program to calculate pressure drop.

12. Using the `if-else` construction, write a program to evaluate the answers to a series of questions and determine if a student will be in the first, middle, or last third of an alphabetical grouping.

13. Cars are usually manufactured with 13-, 14-, 15-, or 16-inch wheels. We can classify cars by type where A type uses 13-inch, B type 14-inch, C type 15-inch, and D type 16-inch wheels. Write a program that reads the car type and prints the wheel size.

14. Cars with four-cylinder engines only have 13- or 14-inch wheels, whereas cars with six cylinders have 13-, 14-, or 15-inch wheels. Cars with eight cylinders 15- or 16-inch wheels. In order to track engine and transmission performance, cars will be separated into seven groups by engine and wheel size. Write a program that reads engine type and wheel size from standard input and displays the category into which this vehicle will be placed.

15. Implementation of total quality management (TQM) requires that manufacturers measure how well parts comply with specification and reject parts that are out of specification. Assume a manufactured product has five key areas of specification: paint, fittings, welds, final assembly, and subassemblies. If a part scores out of compliance on two of the five areas, then the part is rejected. Write a program that will evaluate the input from an operator on each of these qualities as a score from 1 to 10, where 10 is fully acceptable. A score in an area less than 8 is considered to be out of compliance for that attribute. Thus, for example, a product might have the following scores:

paint = 9
fittings = 8
welds = 7
final assembly = 8
subassemblies = 6

Design the program to print a rejection or acceptance message, depending on the number of attributes out of specification. This example would be rejected. (*Note:* These are much less stringent standards than those actually used, but the resulting program works the same.)

6 Controlling the Flow Through Repetition

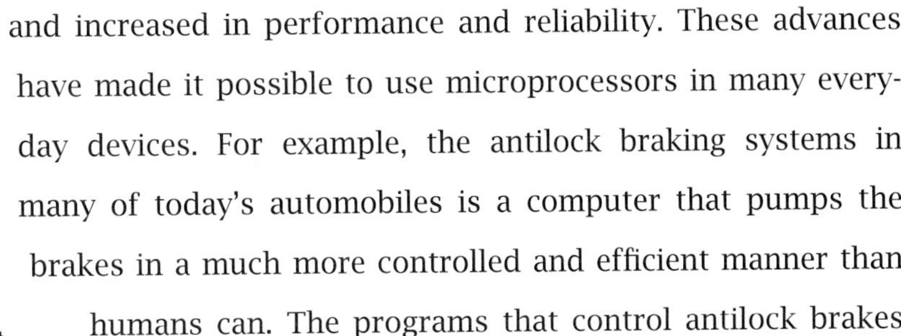

Microprocessor The microprocessor is an ultracompact central processing unit that makes up the "brains" of a digital computer. Since the mid-1980s, microprocessors have dramatically reduced in size and increased in performance and reliability. These advances have made it possible to use microprocessors in many everyday devices. For example, the antilock braking systems in many of today's automobiles is a computer that pumps the brakes in a much more controlled and efficient manner than humans can. The programs that control antilock brakes must repetitively monitor brake sensors.

INTRODUCTION

In Chapter 5 you learned about controlling the program's flow through selection statements. You also learned how to build the conditional expressions that cause a selection statement to behave the way you want it to. In this chapter we extend the idea of directing the flow of control by introducing you to repetition or looping.

Many programming problems repeat the same sequence of actions more than once. Rather than rewrite these actions multiple times, *C* provides you with programming tools that cause a code segment to execute repeatedly. To accomplish this repetition you will learn about the `while` loop, the `do-while` loop, and the `for` loop.

Repetition takes two basic forms. *Conditional looping* involves the repetition of a sequence of instructions for as long as a certain condition holds. *Iterative looping* involves the repetition of a sequence of instructions some predetermined number of times. The `while` and `do-while` loops are conditional loops, whereas the `for` loop is an iterative loop. You will learn how to choose the most appropriate looping structure for your task.

6-1 REPEATING ACTIONS WITH THE `while` LOOP

The `while` loop is called a *pre-test* conditional loop because it tests its loop control condition prior to entering the loop. As long as the condition remains true, the `while` will continue to iterate. Each cycle through the body of a loop is called an *iteration.* The `while` loop takes the following syntactical form:

```
while (<condition>)
<code block>
```

In structure, the `while` loop looks very much like the simple `if` statement, and in some ways they are similar. Like the simple `if` statement the `while` loop consists of a header and a code block or body. The simple `if` statement can be read, "if the condition is true, then execute the statements in the code block." The `while` loop can be read, "while the condition is true, keep executing the statements in the code block." In the case of the simple `if`, the code block is executed at least zero times and at most one time. In the case of the `while` loop, if the condition is initially false, the code block is not executed. Conversely, the `while` loop is theoretically unlimited in the maximum number of times it will be executed.

Recall the space shuttle application at the end of Chapter 3 that calculates vertical and horizontal distances at some time after lift-off. This program would be more interesting if it printed a table of distances for every 5 seconds during the first 2 minutes after lift-off. Such a table would give the user a clearer picture of the shuttle's escape path. The following code segment includes the lines from the Chapter 3 program that calculate vertical and horizontal distance traveled at `time` seconds after lift-off.

```
/* Calculate vertical and horizontal distance */
vertical_distance = (initial_vert_velo * time) -
                    (16 * pow(time,2));
horizontal_distance = initial_hor_velo * time;
```

Modifying this program as described requires that these statements execute 24 times where the value of time ranges from 5 to 120 in increments of 5. You can use a while loop to repeat these statements. The body of the loop will contain these statements together with a statement to print the distances just calculated. The loop should continue iterating as long as the value of time is less than or equal to 120. Example 6-1 contains a program that uses a while loop to implement this idea. The while loop is denoted in boldface type.

EXAMPLE 6-1   A while Loop for Printing Shuttle Distances

```
/*------------------------------------------------------------------
  Description: This program calculates and prints a table of
      the horizontal and vertical distances traveled by a
      space shuttle every 5 seconds for the first 2 minutes
      after launch given initial vertical and horizontal
      velocities in feet per second. The program uses the
      formulas:

          VertDistance = InitVertVelocity(Time) - 16(time)^2

          HorizontalDistance = InitHorizVelocity(Time)

  Programmer: Ken Collier
  Date of Last Revision: 1-1-1995
  Modifications
          Date            Description
          none            none
--------------------------------------------------------------------*/
#include <stdio.h>
#include <math.h>

void main(void)
{
  int     time;                  /* Time elapsed since lift-off */
  double  initial_vert_velo,     /* Initial vertical velocity */
          initial_hor_velo,      /* Initial horizontal velocity */
          vertical_distance,     /* Vertical distance traveled */
          horizontal_distance;   /* Horizontal distance traveled */

  /* Read initial horizontal and vertical velocity */
  printf("Enter initial vertical velocity (feet/sec): ");
  scanf("%lf", &initial_vert_velo);
  printf("Enter initial horizontal velocity (feet/sec): ");
  scanf("%lf", &initial_hor_velo);

  /* Print table header */
  printf("\nTime\tHorizontal Dist.\tVertical Dist.\n\n");

  time = 5;                      /* Initialize LCV */
```

```
   while (time <= 120){    /* Test LCV */
      vertical_distance = (initial_vert_velo * time) -
                          (16 * pow(time,2));
      horizontal_distance = initial_hor_velo * time;
      printf("%d\t%f\t\t%f\n",
             time, horizontal_distance, vertical_distance);
      time += 5;                 /* Update LCV */
   } /* end while */
}
```

. . . .

A sample execution of this program produces the output

```
Enter initial vertical velocity (feet/sec):
2000
Enter initial horizontal velocity (feet/sec): 5

Time      Horizontal Dist.        Vertical Dist.

5         25.000000               9600.000000
10        50.000000               18400.000000
15        75.000000               26400.000000
20        100.000000              33600.000000
25        125.000000              40000.000000
30        50.000000               45600.000000
35        175.000000              50400.000000
40        200.000000              54400.000000
45        225.000000              57600.000000
50        250.000000              60000.000000
55        275.000000              61600.000000
60        300.000000              62400.000000
65        325.000000              62400.000000
70        350.000000              61600.000000
75        375.000000              60000.000000
80        400.000000              57600.000000
85        425.000000              54400.000000
90        450.000000              50400.000000
95        475.000000              45600.000000
100       500.000000              40000.000000
105       525.000000              33600.000000
110       550.000000              26400.000000
115       575.000000              18400.000000
120       600.000000              9600.000000
```

You should notice several things about this loop. First, notice that the expressions for calculating vertical and horizontal distance are unchanged —they have simply become the body of the loop. Second, notice that there are three references to the variable `time`, not including the calculations. In this loop, `time` is the *loop-control variable* (LCV). Every loop has an LCV that is used to determine how many times the loop is repeated. An LCV must undergo three phases in order for a loop to work properly.

1. The LCV must be initialized. This initialization is generally performed before the loop and is executed only one time. In Example 6-1 the statement

```
time = 5;
```

   initializes the LCV to the value 1.
2. The LCV must be tested in the conditional expression to determine whether the loop is to repeat. In Example 6-1, the conditional expression

```
(time <= 120)
```

   combined with the initialization causes the loop to begin executing.
3. The LCV must be updated within the loop. The statement

```
time += 5;
```

   causes the LCV to be incremented by 5 during each repetition of the loop.

Because each of these phases exists in the example, `time` will eventually contain the value 125, causing the loop's condition to be false and thereby causing the loop to terminate. Without the LCV update, the loop will continue forever with the same value for `time`. Without the LCV test, the loop will continue forever because it is never told to stop. This problem is known as an *infinite loop* and should be avoided at all costs.

The code block of a `while` loop has the same flexibility and limitations of the code block of an `if` statement. It can be simple or compound, and it may contain almost any sequence of executable statements that may appear elsewhere in a program, including other `while` loops.

In Example 5-6 you learned how to use a simple `if` statement to develop the following code for detecting user input errors. You saw that although this would detect one error, it would not detect an error on the second data entry. You can use a `while` loop to write a more robust error-detection scheme. Such a `while` loop continues repeating the error message for as long as the user continues to enter invalid data. Example 6-2 shows this revision to the program in Example 5-6.

EXAMPLE 6-2   A `while` Loop for Error Detection

```
/*-------------------------------------------------------------
   Description: This program will print the year the user was born
      given the current year and the user's age as inputs.
   Programmer: Ken Collier
   Date of Last Revision: 1-1-1995
   Modifications
         Date                 Description
         none                 none
-----------------------------------------------------------*/
#include <stdio.h>
```

```
void main(void)
{
  int year, age;

  /* Read age */
  printf("Enter your age followed by the current year: ");
  scanf(" %d%d", &age, &year);

  /* Check for error using while */
  while ((age < 0) || (age > 120)){
    printf("Error - invalid age - reenter: ");
    scanf(" %d", &age);
  } /* end while */

  /* Print birth year */
  printf("You were born in %d.\n", year-age);
}
```

. . . .

Given a sequence of invalid ages, the output of this program is

```
Enter your age followed by the current year:
150 1995
Error - invalid age - reenter: 130
Error - invalid age - reenter: 127
Error - invalid age - reenter: 123
Error - invalid age - reenter: 120
You were born in 1875.
```

Notice that the only difference between the two programs is that the keyword if was switched to the keyword while. In this example the LCV is age. Notice that age is initialized in the first call to scanf(), tested in the conditional expression, and updated inside the loop by the second call to scanf(). But, unlike the earlier loop example, which was predetermined to repeat 120 times, the number of repetitions of this loop is unpredictable. If the user enters a valid age on the first call to scanf(), the loop will not execute even once. On the other hand, the loop will repeat for as long as the user continues entering invalid data. It is possible, though unlikely, that the user may do this forever. This does not mean that we have created an infinite loop. Because the LCV is updated inside the loop, it is given the opportunity to cause the loop condition to become false on each iteration. An infinite loop is one whose condition does not have the opportunity to become false and is guaranteed to continue repeating once it has begun.

The previous two examples are conceptually quite different. As noted, Example 6-1 is predetermined to iterate 120 times, wheras the loop in Example 6-2 is unpredictable. The first loop is an example of an iterative loop, and the second is an example of a conditional loop. As we will see, there is a better way to write the iterative loop using the for construct in C.

As a final example, we revisit the program for calculating the average of a series of user-entered numbers presented in Chapter 1. In this version, the statements have been numbered to help with a walk-through of the program, these numbers are not part of the program.

EXAMPLE 6-3   A Program for Calculating Averages

```c
#include <stdio.h>

void main(void)
{
(1)   float number,      /* Temporary storage of each new value */
(2)       average,       /* Stores average */
(3)       total=0;       /* Running total */
(4)   int count=0;       /* Number of numbers */
(5)
(6)   /* Get the first number */
(7)   printf("Enter the next data value (0 to quit data entry): ");
(8)   scanf("%f", &number);
(9)
(10)  /* Calculate total of all numbers entered and count the numbers */
(11)    while (number != 0.0){
(12)    total += number;
(13)    count += 1;
(14)    printf("Enter the next data value (0 to quit data entry): ");
(15)    scanf("%f", &number);
(16)  }
(17)
(18)  /* Calculate and print the average */
(19)  /* First check to make sure that count is not 0 */
(20)  if (count != 0){
(21)    average = total/count;
(22)    printf("The average of your data is %f.\n", average);
(23)  }
(24)  else
(25)    printf("No data entered.\n");
    }
```

. . . .

Notice that we have made a few changes in this program from the one introduced in Chapter 1. The most important of these is the addition of an if statement to ensure that we do not divide by 0 when calculating the average. This is another example of the type of error checking you should practice in your own programs.

Recall that this program reads the user's input and adds it into a running total until the user enters a value of 0. Additionally, the program counts the number of numbers entered by the user and stores this number in the variable count. After exiting from the while loop, the average is calculated by dividing the value in total by the value in count. To help you understand what is happening in this program, we introduce the table trace.

A *table trace* is a hand execution of a computer program to help you understand all the subtle behaviors of the program. You can use it to help you find the errors in your programs. To conduct a table trace, you build a table that contains a column for each variable in the program and a column for the output. The program's statements are numbered for reference, and a column is added to the table for these line numbers. Once the table is built, you execute the program by hand and record the changing values of the

Table 6-1   Table Trace of Average Calculator

Statement	Number	Average	Total	Count	Output
1, 2			0.0	0	
8	5.0				
12			5.0		
13				1	
15	3.0				
12			8.0		
13				2	
15	10.0				
12			18.0		
13				3	
15	8.0				
12			26.0		
13				4	
15	7.0				
12			33.0		
13				5	
15	0.0				
21		33.0/5 = 6.6			
22					6.600000

variables in the appropriate columns. Table 6-1 shows a table trace of the average calculating program given the input values 5.0, 3.0, 10.0, 8.0, 7.0, and 0.0.

The rows set apart by bold lines represent those statements within the while loop that were executed repeatedly until the value 0.0 was entered for number, causing the while condition to fail and the loop to terminate. To ensure you understand the behavior of the loop, the concept of a running total, and the concept of a counter variable, you should trace this execution by hand. These programming concepts have a place in many different programs, and it is important for you to understand how they work.

Try It

Perform a table trace on the average calculator program given the inputs 3.5, 19.2, 14.3, 5.0, 3.333, and 0.0.

What If

Write a program to calculate the series $\sum_{i=1}^{n} i^2 = 1^2 + 2^2 + 3^2 + \cdots + n^2$ using a while loop, where n is entered by the user from standard input.

6-2   REPEATING ACTIONS WITH THE do-while LOOP

The while loop tests the truth of its condition before entering the loop's body. In certain cases it is better to test the condition at the end of the loop. The do-while loop is a *post-test* conditional loop that performs its test at the end of the loop. This loop has the following general form:

```
do
   <code block>
while(<condition>);
```

As noted, the `while` loop is not guaranteed to execute at all. Because the loop test appears at the end of the loop, the `do-while` loop is guaranteed to execute at least once.

The `do-while` loop can be read, "Execute the statements in the code block for as long as (while) the condition remains true." The `do-while` loop is better than the `while` loop for handling input errors. Using the `while` loop to detect errors, it is necessary to prompt and read once prior to the loop to initialize the LCV and then again within the loop to update the LCV.

```c
printf("Enter your age: ");
scanf("%d", &age);
while ((age < 0) || (age > 120)){
   printf("Error - value invalid reenter: ");
   scanf("%d", &age);
}
```

The minor difference in these two calls to `printf()` and `scanf()` is in the wording of the prompt. A little rewriting combined with the replacement of `while` with `do-while` allows us to eliminate the duplicate calls to `printf()` and `scanf()`. Because the `do-while` loop is guaranteed to execute at least once, it can be used for both initialization and updating.

```c
do{
   printf("Enter your age (must be in the range 0-120): ");
   scanf("%d", &age);
} while ((age < 0) || (age > 120));
```

Note that the LCV still goes through the three phases of initialization, testing, and update within the loop. On the first iteration of the loop, age is initialized. It is then tested at the end of the loop. On successive iterations, age is updated. Hence, initialization and update are handled by the same statement.

Try It

Modify the program in Example 6-2 to use a `do-while` loop instead of a `while` loop for error detection.

6-3 REPEATING ACTIONS WITH THE for LOOP

Sometimes it is possible to specify the number of times you wish to repeat a set of statements. The `for` loop is more appropriate for such cases. The `for` loop is an iterative loop and is used whenever the number of loop iterations is predetermined, either by the programmer or by some input or calculated value.

As with the `while` and `do-while` loops, the `for` loop must have at least one LCV that is initialized, tested, and updated. However, the `for` loop handles each of these phases quite differently from the other loops. The general form of the `for` loop is

```
for ([<initialize lcv>]; [<test condition>]; [<update lcv>])
    <code block>
```

Notice that the three phases of the LCV are contained in the header of the `for` loop. This allows the code block to contain only the statements that pertain to solving the problem.

The code block of a `for` loop is subject to all the flexibility and limitations of code blocks in the other control structures.

In Example 6-1 you learned how to use a `while` loop to print a table of distances traveled by the space shuttle during the first 2 minutes of flight. Although the `while` loop works fine for this problem, the `for` loop is a better choice because we know beforehand that the loop is to iterate 24 times. Example 6-4 shows the same segment of code written using a `for` loop instead of a `while` loop.

EXAMPLE 6-4   A `for` Loop for Printing Shuttle Distances

```
for (time = 1; time <= 120; time += 5){
   vertical_distance = (initial_vert_velo * time) -
                       (16 * pow(time,2));
   horizontal_distance = initial_hor_velo * time;
   printf("%d\t%f\t\t%f\n",
       time, horizontal_distance, vertical_distance);
} /* End For */
```

. . . .

Note that the only difference in the two implementations is in the physical positioning of the initialization, test, and update of the LCV `time` (and, of course, the change of the keyword `while` to `for`). The `for` loop is slightly more compact because it combines three statements into a single loop header.

Try It

Modify the program in Example 6-1 by replacing the `while` loop with the `for` loop in Example 6-4. Test the program to be sure it still works correctly.

Because the `for` loop is more cryptic than the other loops, it may be helpful to diagram the sequence of events that take place during the execution of a simple `for` loop. Figure 6-1 shows this diagram using a program that prints each integer between 1 and 10 and its square. The arrows reflect the flow of control, and the numbers associated with each arrow show the order of control transition.

Notice from the diagram that LCV initialization is evaluated first and only once, which is consistent with what you saw when using the `while` loop. Immediately following LCV initialization, the loop's condition is evaluated. Again, this behavior is similar to that of a `while` loop. Moreover, like the

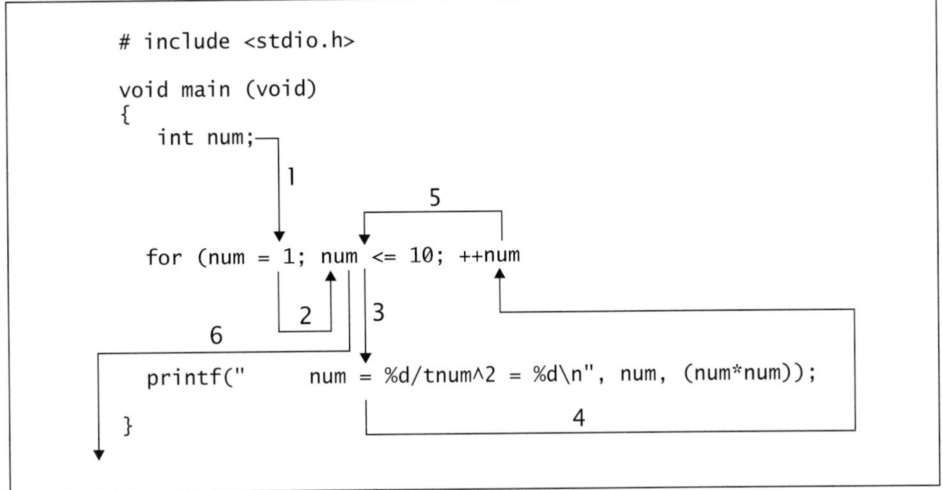

Figure 6-1 The Sequence of Execution in a for Loop

while loop, the LCV is updated, and the condition is reevaluated on each iteration of the for loop. As soon as the condition evaluates to false, the flow of control branches to the statement following the loop's code block (following path number 6 in the diagram). In this case there is no such statement, so the program terminates normally.

The most typical use of the for loop involves an LCV of type int and is incremented by 1 for each iteration and eventually approaches the upper bound described by the conditional expression. However, the for loop is not restricted to this format. Another common use of the for loop is for counting down to a lower-bound value rather than up toward an upper-bound value. For example, the factorial of a number n is expressed mathematically as $n!$ (read n factorial). The factorial of n is defined by the expression $n! = n \times (n - 1)!$, where $0! = 1$ by definition. For example, $5! = 5 \times 4 \times 3 \times 2 \times 1 = 120$. Factorials are used by industrial engineers in the area of probability and statistics. You can use a for loop to calculate the factorial of a number entered by the user.

EXAMPLE 6-5   Using a for Loop to Calculate Factorial

```c
#include <stdio.h>

void main(void)
{
  int number, counter, factorial = 1;

  printf("Enter a positive number: ");
  scanf("%d", &number);

  for(counter = number; counter > 0; --counter)
    factorial *= counter;

  printf("\tThe factorial of %d is %d.\n", number, factorial);
}
```

. . . .

The output from several executions of this program is

```
Enter a positive number: 5
  The factorial of 5 is 120.
Enter a positive number: 6
  The factorial of 6 is 720.
Enter a positive number: 7
  The factorial of 7 is 5040.
Enter a positive number: 8
  The factorial of 8 is 40320.
Enter a positive number: 9
  The factorial of 9 is 362880.
Enter a positive number: 10
  The factorial of 10 is 3628800.
Enter a positive number: 1
  The factorial of 1 is 1.
Enter a positive number: 0
  The factorial of 0 is 1.
```

The first several executions show you that factorials get very large very quickly. The last two executions show that the program works for the value 1 and the special case 0.

You should notice several things about this program. First, notice that the initialization of the LCV `counter` is not a constant as in earlier examples. Although we stated previously that the `for` loop should be used when the number of iterations is predetermined, in this case we say that the number of iterations is symbolically predetermined. That is, although you do not know the actual value of `number` before program execution, you can still be certain that the loop will iterate `number` times as soon as `number` has a value.

The second thing to notice is that the LCV is decremented rather than incremented as in previous examples. This update coincides with the loop's conditional expression, which says that the loop is to continue repeating as long as `counter` remains *greater* than 0.

Finally, let us take a bit of a side-track from the topic of control structures. Observe the similarity between the calculation of the factorial of *n* and the calculation of the running total in the average calculator. Inside a loop, the statement

```
factorial *= counter;
```

can be thought of as a "running product" with a similar structure to the running total expression

```
total += number;
```

This expression structure has a variety of applications. For either running total or running product to work, the running variable (`factorial` or `total`) must have the proper initial value. In the previous examples, `factorial` is initialized to 1, and `total` is initialized to 0. This difference in initial values stems from the arithmetic identities: $0 + n = n$, and $1 \times n = n$. You can take advantage of these properties to ensure that the initial value does not affect the actual data being added or multiplied into the running variable.

Although it is rare, the `for` loop can be used in some very nontraditional ways. Note that the square brackets in the general form description of the `for` loop tell you that the contents of each field in the `for` loop header are optional. These options mean it is valid to write a `for` loop like

```
for (;;)
    printf("Infinite loop");
```

In this case there is no LCV, and the loop has no terminating condition; therefore, this is an infinite loop. Although this code is not very useful, it is possible to write a meaningful `for` loop that omits some of the fields. Consider the following example:

```
printf("Enter your age: ");
scanf("%d", &age);
for ( ;(age < 0) || (age > 120); ){
    printf("Error - value invalid reenter: ");
    scanf("%d", &age);
}
```

Does this code look familiar? This is essentially the `while` loop we developed earlier for detecting user input errors. We have simply moved the initialization and update of the LCV, age, out of the `for` loop header and into their traditional positions as if this were a `while` loop. Although this is of no benefit and is a misuse of the `for` loop, it is helpful in developing a better understanding of the way `for` loops work. All *C* cares about is that the middle field in the `for` loop header is either empty or evaluates to an integer value. If that value is 0 (false), then the loop terminates; otherwise, the loop continues to iterate.

Because the initialization, test, and update of the LCV in a `for` loop are all arithmetic expressions using the *C* operators, the variety of expressions that can occur in the fields of the `for` loop is only as limited as any expression in *C* (which is not very limited). Suppose you wish to write a program to print the values between 1 and 10 in ascending order and 10 and 1 in descending order, side by side. You could write the `for` loop shown in Example 6-6 to accomplish this task.

EXAMPLE 6-6   ## The Power of the `for` Loop

```
#include <stdio.h>

void main(void)
{
    int up, down;

    for (up = 1, down = 10; up <= 10; ++up, --down)
        printf("\t%d\t%d\n", up, down);
}
```

. . . .

The output of this program is

```
 1   10
 2    9
 3    8
 4    7
 5    6
 6    5
 7    4
 8    3
 9    2
10    1
```

An operator that was not discussed in Chapter 3 is the comma operator. The *comma operator* simply allows multiple, independent expressions to appear in the same statement. It is a binary operator and has lower precedence than all other operators. The comma's operands are evaluated from left to right, and a comma expression returns the value of its right operand.

In Example 6-6, the comma allows us to make use of multiple LCVs in the for loop. Although the first and last fields in the for loop header appear to be made up of two expressions, as far as *C* is concerned each is a single expression. We could have used the comma to combine conditional expressions as well with the statement

```
up <= 10, down >= 1
```

However, this statement is not necessary in the previous program because both conditions become false at the same time. Many other nontraditional examples of the for loop could be presented, but most of the time you will not need to take advantage of this flexibility that *C* affords.

Anything you can do with a for loop can also be done using a while loop. Moreover, a while loop can be written to replace any do-while loop. These control structures provide some specialized tools to add to the programmer's toolbox. A good programmer will develop an understanding of the subtle differences of each loop and will select the most appropriate loop for the task at hand. In general, whenever the number of iterations of a loop is known or can be described symbolically before the program is executed, the programmer should choose the for loop. Whenever the task requires that the code block be executed one or more times, it is appropriate to use a do-while loop. Whenever the code in the code block may be executed zero or more times, a while loop is called for.

What If  Write a program that will calculate $\sum_{i=1}^{n} 2^i = 2^1 + 2^3 + \cdots + 2^n$ using a for loop, where n is entered by the user.

6-4 COMBINING CONTROL STRUCTURES

As noted, selection statements can be combined with one another to form complex branching structures. Similarly, repetition statements can appear in the bodies of other repetition statements to form more complex looping structures. Selection statements can also appear in the bodies of repetition statements and vice versa. When one control structure is placed in the body of another control structure, we say that we are *nesting* control structures.

Nested looping structures will become particularly useful when you learn about two-dimensional arrays in Chapter 7. For now, suppose you want to write a program to generate a multiplication table for the product of 0×0 up to 9×9. You can use the following for loop to generate a single row in the table:

```
for (factor2 = 0; factor2 <= 9; ++factor2)
  printf("%5d", factor1 * factor2);
putchar('\n');
```

The 5 in front of the format-conversion code causes the product to print in a field of five spaces. This format-control specifier will help make the output readable. The call to putchar() is not part of the body of the for loop. It prints a new line after the entire row has been printed.

In this loop we refer to one of the factors as factor1. This factor represents the row being generated. For row 0, factor1 is 0; for row 1, factor1 is 1; and so on. The changing values of factor1 suggest that we can use another for loop for advancing the value of factor1 from 0 up to 9. If we place the factor2 loop inside the body of the factor1 loop, we will get 9 rows of 9 products, which describes a multiplication table. Example 6-7 contains a program to implement this nested loop.

EXAMPLE 6-7   Nesting for Loops to Print a Multiplication Table

```
/*-----------------------------------------------------------------
  Description: This program will display a multiplication table
      for the first 81 products.
  Programmer: Ken Collier
  Date of Last Revision: 1-1-1995
  Modifications
          Date                Description
          none                none
------------------------------------------------------------------*/
#include <stdio.h>

void main(void)
{
  int factor1, factor2;    /* LCVs and factors in the table */

  for (factor1=0; factor1 <= 9; ++factor1){
    for (factor2=0; factor2 <= 9; ++factor2)
      printf("%5d", factor1*factor2);
    putchar('\n');
  } /* End For */
}
```

The output of this program is

```
0    0    0    0    0    0    0    0    0    0
0    1    2    3    4    5    6    7    8    9
0    2    4    6    8    10   12   14   16   18
0    3    6    9    12   15   18   21   24   27
0    4    8    12   16   20   24   28   32   36
0    5    10   15   20   25   30   35   40   45
0    6    12   18   24   30   36   42   48   54
0    7    14   21   28   35   42   49   56   63
0    8    16   24   32   40   48   56   64   72
0    9    18   27   36   45   54   63   72   81
```

Although the ways in which control structures can be embedded within one another are almost unlimited, there is one basic rule for doing so. Whenever placing a control structure within the code block of another control structure, the code block of the embedded structure must begin and end within the code block of the outer control structure. So, if you wish to place an if statement within a while loop, the code block of the if statement must be closed before the closing bracket of the while loop. In general, the form of one control structure embedded within another is as follows:

```
<control structure header 1>{
      .
      .
      .
      <control structure header 2>{
            .
            .
            .
      }          /* End of control structure 2 */
}     /* End of control structure 1 */
```

This rule should not be difficult to understand, because it is analogous to the mathematical rules for parenthesizing an expression. Innermost parentheses must open and close within their outer parentheses. Beyond this basic restriction, you are limited only by your creativity and programming ability.

What If

Modify the multiplication table program so that the columns and rows are labeled with the factors corresponding to the product in the table.

Application 1 SIMULATING DIGITAL CIRCUITS

Electrical and Computer Engineering

Digital circuits are used to perform the arithmetic calculations within a computer. Circuits are built from logic gates. Figure 6-2 shows the logic gates necessary to build a circuit.

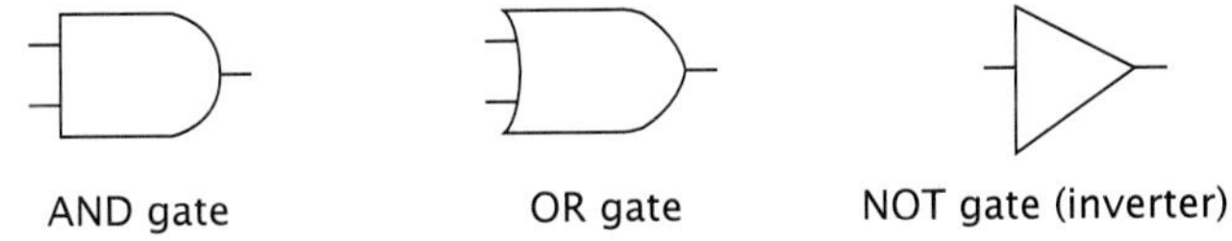

Figure 6-2 Logic Gates

Inputs to these gates are the presence or absence of electrical pulses along the lines leading into the gate, and outputs are pulses along the lines leading out of the gates. The output line of the AND gate contains a pulse only if both input lines contain a pulse. The output line of the OR gate contains a pulse if either or both of the input lines contain a pulse. The output line of the NOT gate contains a pulse when a pulse is absent on the input line, and the output line does not contain a pulse if the input line contains a pulse.

By combining these gates, electrical and computer engineers are able to build circuits to perform arithmetic and other functions. For example, a circuit called a half-adder is used to add two binary digits together. Figure 6-3 shows the logical diagram for a half-adder circuit.

This circuit can be described by the logical expression

$$!(X \;\&\&\; Y) \;\&\&\; (X \;||\; Y)$$

The logic for the carry value is

$$X \;\&\&\; Y$$

You can use a truth table to convince yourself that the circuit's outputs are correct for all values of X and Y. We will use 1 to represent an electrical pulse and 0 to represent the absence of a pulse. Table 6-2 is a truth table showing all possible input values for X and Y and the corresponding output values of Carry and Sum. A simple hand calculation of these binary additions reveals that the values on the Sum and Carry lines are correct.

When adding binary numbers with multiple digits, you must add in any carry bits that come in from the right of the place value currently being added and send to the left any carry values resulting from the addition. A full-adder circuit is a combination of half-adders that is designed to ac-

Table 6-2   Truth Table for Half-Adder

X	Y	Carry	Sum
0	0	0	0
0	1	0	1
1	0	0	1
1	1	1	0

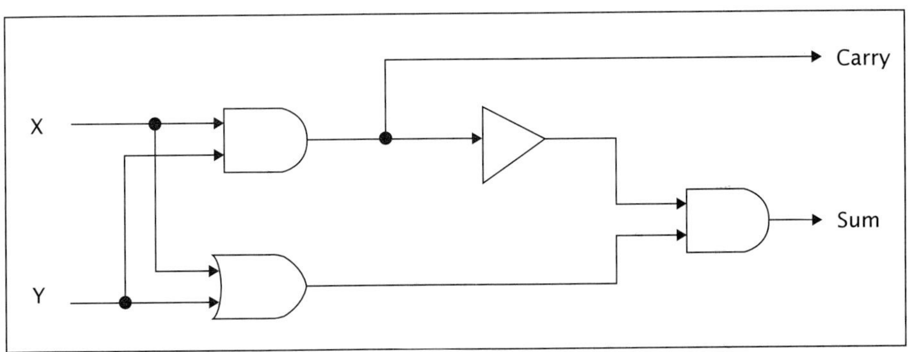

Figure 6-3 Circuit Diagram for a Half-Adder

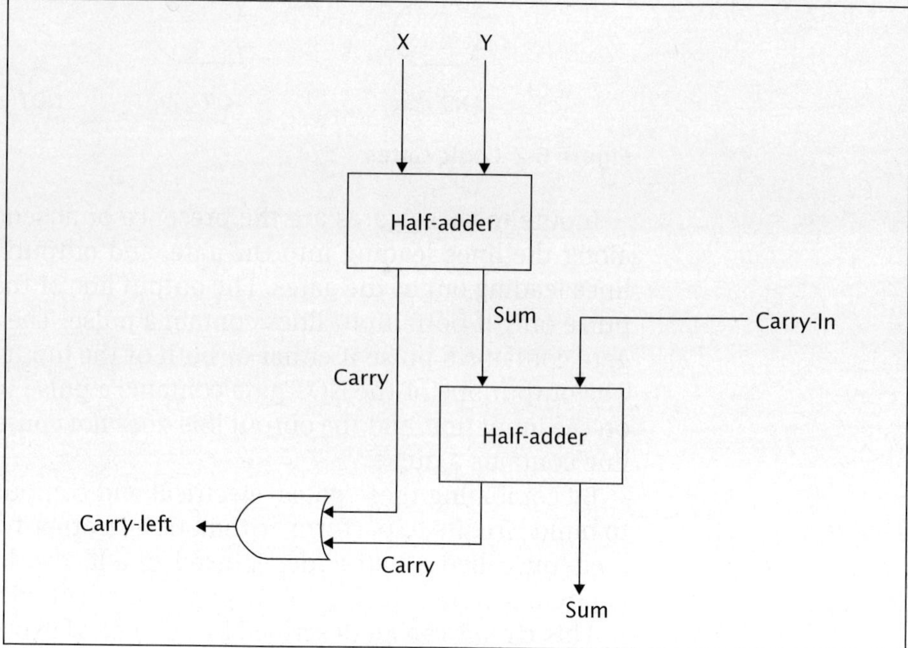

Figure 6-4 Circuit Diagram for a Full-Adder

cept as input the carry value from a previous binary addition (presumably from the binary digits to the right), add the carry value in, and send any resulting carry out. Figure 6-4 shows the logic diagram for a full-adder.

The logic for a full-adder is expressed by the logical expression

```
!((!(X && Y) && (X || Y) && Carry-in) &&
  ((!(X && Y) && (X||Y) || Carry-in)
```

and the logic for the Carry-left value is

```
(X && Y) || ((!(X && Y) && (X || Y)) && Carry-in)
```

Using this logic, you can write a *C* function to simulate a full-adder. Then, using this function, you can write programs to perform addition on binary values of multiple digits.

Write a program that will read two 8-bit binary numbers from standard input, add them together using a full-adder function, and display the sum on standard output.

1. Define the Problem

The program must read two 8-bit binary numbers from standard input, add them together using binary addition, and display the result on standard output. One of the difficulties in this problem will be reading a binary number, because *C* does not directly provide a data type for this purpose. Because you have been asked to use a full-adder to perform the addition, you will also have to peel a digit at a time from both of the binary numbers entered by the user. Once these digits have been added, the result must be placed in its proper place in the final sum, and the carry value must be available to add into the next two digits.

2. Gather Information

The input data required is two binary numbers. You will actually read these values in as decimal integers but treat them as if they were binary values. Any digits in the input values that are not 0 or 1 are considered errors and must be handled. Numbers larger than eight digits must not be accepted. Errors should be handled using a do-while loop.

Output from the program is the result of binary addition on the two input values. This output will be printed as a decimal integer containing only 0s and 1s. There is the additional possibility of a final carry bit. This value will also be printed on standard output.

3. Generate and Evaluate Potential Solutions

This solution requires you to apply the full-adder to the bits in each place value one at a time. Therefore, a looping structure will be used to move through the user-entered numbers from right to left. For each place value, you must get the next rightmost digit, add the digits using a full-adder, retain the carry value, and place the sum into its proper place in the final result. These steps are described by the following algorithm:

```
For each digit in two 8 digit numbers
   Get the least significant bit from the first number
   Get the least significant bit from the second number
   Add them together using a full-adder
      Retain the carry bit
   Place the sum in its proper place in the final result
End loop
```

This driver algorithm suggests that you need several functions to help you out. First, you need a function that will read a binary number and check for errors. It will be useful to have a function that strips off the rightmost digit of an integer and returns the digit and the number minus its rightmost digit. Of course, a full-adder function will be necessary. This function might make calls to a half-adder function to perform its purpose. Finally, a function to build up the final result digit by digit will be helpful. Figure 6-5 contains a structure chart for the solution to this problem.

According to the structure chart, this program will be made up of six function definitions. Reading and checking for errors requires some thought. Recall that you must check for two kinds of errors: the 8-bit limit and binary digits only. Earlier you saw the do-while loop used for

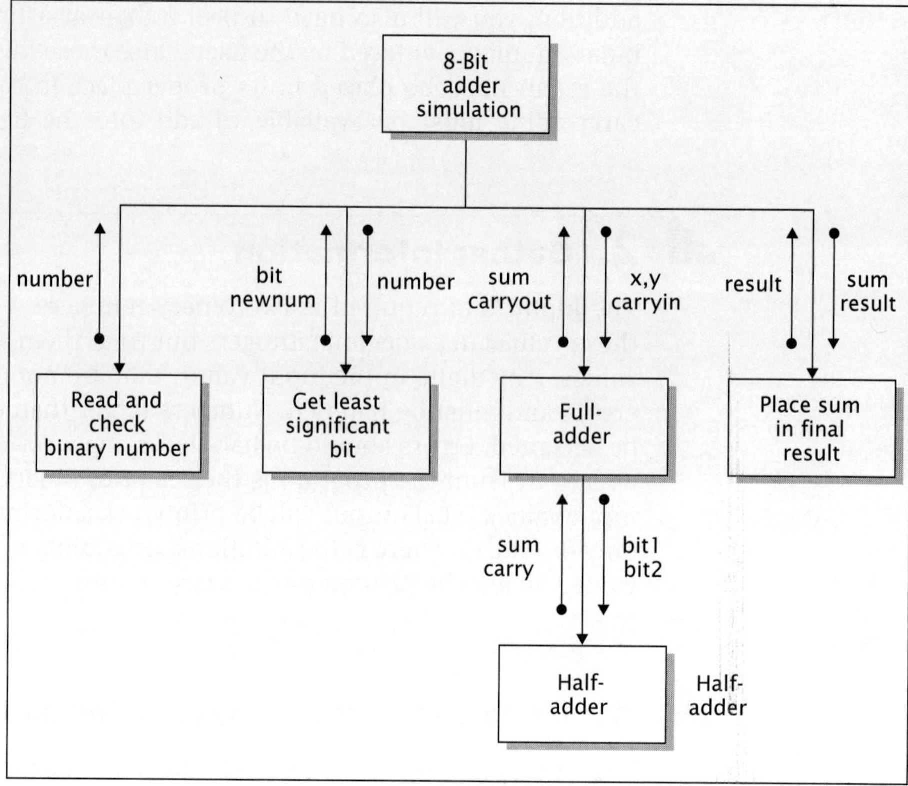

Figure 6-5 Structure Chart for a Binary 8 Bit Adder

error detection. You can use the same construct in this case, but you will need to check for both types of errors within the loop. To do this you can use an error flag. An *error flag* is a variable that is set to 0 (false) initially. If an error is detected, the flag is set to 1 (true). The error flag is used as a loop-control variable. If the flag is true, the loop iterates; otherwise, the loop terminates. The following algorithm describes the logic for reading and checking a binary number.

EXAMPLE 6-8   **Algorithm for the Function `get_binary_number()`**

```
do
   set error flag to 0
   read number
   if number > 11111111 then
     set error flag to true
   for each digit in number
     if rightmost digit is not a 1 or 0
       set error flag to 1
   if error flag is 1
     print error message
while error flag is 1
```

Notice that the body of this do-while loop contains several other control structures. One of these is a for loop to check each digit for validity. To implement this for loop, you can use the function that returns the least-significant digit of a binary number. To implement that function definition, you can make use of the % operator and the / operator. The following algorithm describes the logic of this function.

EXAMPLE 6-9

Algorithm for the Function `get_least_significant_bit()`

```
set least_sig_bit to number % 10
set number to number/10
```

The first step in this algorithm extracts the rightmost digit of the number, whereas the second step modifies the number by removing its rightmost digit.

The full-adder function will make two calls to the half-adder function. Example 6-10 shows the algorithm for the half-adder function, and Example 6-11 contains the algorithm for the full-adder function.

EXAMPLE 6-10

Algorithm for the Function `half_adder()`

```
if x and y then
   set carry to 1
else
   set carry to 0
if not(x and y) and (x or y) then
   return 1
else
   return 0
```

This algorithm corresponds to the logic described earlier.

EXAMPLE 6-11

Algorithm for the Function `full_adder()`

```
set intermediate_sum to half_adder(x,y,carry1)
set final_sum to half_adder(intermediate_sum, carryin, carry2)
if carry1 or carry2
   set carryout to 1
else
   set carryout to 0
return final_sum
```

Trace both of these functions, and compare them to the logic diagrams presented previously to convince yourself that they function the same way.

The last function you must define is the insertion of the result of the full-adder into its proper place value in the result. When all bits have been processed, the result will contain the binary sum of the addition process. Example 6-12 contains this algorithm.

EXAMPLE 6-12

Algorithm for Function `add_to_final_sum()`

```
number = (number + (place_value * bit))
return (number)
```

. . . .

For this algorithm to be effective, the variable `place_value` must be updated for each new bit. You will use `place_value` as the LCV of the `for` loop in `main()` and update it by multiplying it times 10 on each iteration to get the proper place values.

4. Refine and Implement a Solution

The structure chart and algorithms developed in the previous section are implemented as the program in Example 6-13.

EXAMPLE 6-13

Simulating a Full-Adder to Add Two 8-Bit Numbers

```c
/*-------------------------------------------------------------
  Description: This program simulates the logic for a set of
     full-adders designed to add two 8-bit binary numbers.
  Programmer: Ken Collier
  Date of Last Revision: 1-1-1995
  Modifications
       Date              Description
       none              none
-------------------------------------------------------------*/
#include <stdio.h>

int get_binary_number(void);
int get_least_sig_bit(int number, int *bit);
int half_adder(int x, int y, int *carry);
int full_adder(int x, int y, int carryin, int *carryout);
int add_to_final_sum(int number, int bit, int place_value);

void main()
```

```c
{
  int num1, num2, /* To store binary operands */
      bit1, bit2, /* To store bits to be added */
      sum,        /* Result of adding bit1 and bit2 */
      carryin=0,  /* Carries value from the right */
      carryout,   /* Carries value to the left */
      result=0,   /* Final result of adding all bit pairs */
      place_val;  /* Loop control variable to hold 1,10,100... */

  /* Read the two binary operands */
  printf("Enter two 8-bit binary numbers\n");
  printf(" First number: ");
  num1 = get_binary_number();
  printf(" Second number: ");
  num2 = get_binary_number();

  /* Now add numbers bit at a time */
  for(place_val = 1; place_val <= 10000000 ; place_val *= 10){
    /* Peel off least significant digits */
    num1 = get_least_sig_bit(num1, &bit1);
    num2 = get_least_sig_bit(num2, &bit2);

    /* Add bits together */
    sum = full_adder(bit1, bit2, carryin, &carryout);

    /* Update the carryin bit for next iteration */
    carryin = carryout;

    /* Put the sum into result */
    result = add_to_final_sum(result, sum, place_val);
  } /* end for */

  /* Print result */
  printf("Sum is: %8d\nCarry bit is: %d\n", result, carryout);
}

/*-------------------------------------------------------------
Module: get_binary_number(void)
Description: Reads a binary number as an int and checks to
  ensure the number is a maximum of 8 bits and contains only 1s
  and 0s.
Input Parms: none
Returns: A valid 8-bit binary number
Modifications
        Date            Description
        none            none
-------------------------------------------------------------*/
int get_binary_number(void)
```

```c
{
  int number,        /* Number from standard input */
      tmp,           /* Copy of number that can be destroyed */
      bitcnt,        /* LCV for checking bit validity */
      bit,           /* A single digit extracted from number */
      error;         /* Error detection flag */

  do{
    error = 0;       /* Assume no input error */
    scanf("%d", &number);

    /* Check to make sure it is an 8 bit number */
    if(number > 11111111)
      error = 1;     /* Set error flag if true */

    /* Check for only digits of 0 and 1 */
    tmp = number;
    for(bitcnt = 0; bitcnt < 8; ++bitcnt){
      tmp = get_least_sig_bit(tmp, &bit);
      if ((bit != 1) && (bit != 0))
        error = 1;/* Set error flag if true */
    } /* End For */

    if (error)
      printf("Error: Must be 8 bit binary number - reenter:");
  }while(error);

  return(number);
}


/*-------------------------------------------------------------
Module: get_least_sig_bit(int number, int *bit)
Description: Extracts the last digit of an integer.
Input Parms: Number to be modified.
Output Parms: Pointer to a storage location for the extracted
  digit.
Returns: The input number minus its rightmost digit.
Modifications
        Date            Description
        none            none
-------------------------------------------------------------*/
int get_least_sig_bit(int number, int *bit)
{
  *bit = number % 10;
  return((int)(number/10));
}


/*-------------------------------------------------------------
Module: half_adder(int x, int y, int *carry)
Description: Simulates a half-adder circuit
Input Parms: Bits to be added.
Output Parms: Pointer to storage location of carry value.
Returns: The sum of x and y
```

```c
Modifications
        Date               Description
        none               none
---------------------------------------------------------------*/
int half_adder(int x, int y, int *carry)
{
  /* First calculate carry */
  if (x && y)
    *carry = 1;
  else
    *carry = 0;

  /* Now return sum */
  if ((!(x && y) && (x || y)))
    return(1);
  else
    return(0);
}


/*---------------------------------------------------------------
Module: full_adder(int x, int y, int carryin, int *carryout)
Description: Simulates a full-adder circuit
Input Parms: Bits to be added, and carryin value.
Output Parms: Pointer to storage location of carryout value.
Returns: The sum of x and y
Modifications
        Date               Description
        none               none
---------------------------------------------------------------*/
int full_adder(int x, int y, int carryin, int *carryout)
{
  int intermediate_sum,  /* Stores sum from first half-adder */
      sum,               /* Stores sum from full-adder */
      carry1, carry2;    /* Stores intermediate carry values */

  /* First half adder */
  intermediate_sum = half_adder(x, y, &carry1);

  /* Second half adder */
  sum = half_adder(intermediate_sum, carryin, &carry2);

  /* Calculate carryout */
  if (carry1 || carry2)
    *carryout = 1;
  else
    *carryout = 0;

  return(sum);
}
```

```
/*-----------------------------------------------------------
Module: add_to_final_sum(int number, int bit, int place_value)
Description: Inserts a bit value into a larger binary number at
   the place value position specified.
Input Parms: Number for bit to be inserted, bit value, place
   value for insertion.
Output Parms: none
Returns: Modified number
Modifications
        Date            Description
        none            none
-----------------------------------------------------------*/
int add_to_final_sum(int number, int bit, int place_value)
{
   return(number + (place_value * bit));
}
```

. . . .

5. Verify and Test the Solution

This program should handle all positive input values. If the value input is
larger than eight digits or contains invalid digits, error trapping will
occur, and the program should behave as required. If the inputs are valid,
the output should be the correct sum of the binary addition of these
values. Additionally, the carry value is output to reflect any overflow. To
test these specifications, you should enter values that are valid, invalid,
and on the boundaries (for example, 1111 1111 and 0000 0000). The
following output shows several test runs of the program, with a variety of
valid and invalid values:

```
Enter two 8-bit binary numbers
   First number: 00000001
   Second number: 00000001
Sum is:        10
Carry bit is: 0

Enter two 8-bit binary numbers
   First number: 10101010
   Second number: 01010101
Sum is: 11111111
Carry bit is: 0

Enter two 8-bit binary numbers
   First number: 12345678
Error: Must be 8 bit binary number – reenter:11111111
   Second number: 11111111
Sum is: 11111110
Carry bit is: 1

Enter two 8-bit binary numbers
   First number: 10011001
   Second number: 01100110
```

```
                    Sum is: 11111111
                    Carry bit is: 0

                    Enter two 8-bit binary numbers
                       First number: 00000000
                       Second number: 00000000
                    Sum is:         0
                    Carry bit is: 0
```

From these results, it appears the program works as specified. Run the program with your own test values to see if it works properly.

Try It   The current version of the circuit simulation program does not trap errors caused by negative inputs. Modify the `get_binary_number()` function so that it handles this type of error.

What If   The circuit simulation program would be more useful if it allowed the user to continue entering binary numbers to add until he or she entered a sentinel value to quit. Modify the program so that after the first two numbers have been added, the user is asked if he or she would like to add two new numbers. Accept Y or N as a response to this prompt, and loop until the user enters N.

SUMMARY

This chapter introduced you to the details of the repetition structures provided by C. You learned how to repeat statements using the `while`, `do-while`, and `for` loops. These control structures each have a header and a body and a loop-control variable. The loop-control variable must always be initialized once, then tested and updated on each loop iteration. Each loop construct serves a slightly different purpose. The `while` and `do-while` loops are conditional loops, and the `for` loop is iterative.

Now that you know how to control the flow in your programs using sequencing, selection, and repetition, you can write very powerful programs to solve engineering problems. In the next chapter you will learn how to use `for` loops to manipulate arrays.

Key Words

comma operator	loop-control variable
conditional looping	nesting
error flag	pre-test loop
infinite loop	post-test loop
iteration	table trace
iterative looping	

Exercises

1. Perform a table trace of the following program. What is printed?

```c
#include <stdio.h>

void main(void)
{
   int counter;

   for(counter=2; counter<=50; counter+=2)
     if (counter % 3 == 0)
       printf("%d\n", counter);
}
```

2. Write a function to match the prototype

```c
int factorial(int number)
```

   that will accept an integer as its argument and will return the factorial of that number.

3. In the problems at the end of Chapter 4 you wrote several functions for converting between temperature scales. Write a program using those functions to print a temperature-conversion table for temperatures between 0°C and 100°C in increments of 5. Use a `while` loop to create your table.

4. The `while` loop used in problem 3 would be more appropriate as a `for` loop. Do you agree with this statement? Why, or why not?

5. Rewrite the `while` loop of problem 3 as a `for` loop.

6. Problem 12 in Chapter 3 presents a better set of formulas for calculating space shuttle distances traveled. Write a program using these formulas that prints a nicely formatted table of vertical and horizontal distances for the first 2 minutes after lift-off on 5-second intervals similar to Example 6-1.

7. Which is the best loop structure for the program in problem 6? Why? If you wrote your program using a `while` loop, rewrite it using a `for` loop. If you wrote your program using a `for` loop, rewrite it using a `while` loop.

8. In problem 12 of Chapter 2 you wrote a program for calculating distance traveled given rate of speed and time elapsed. Write a program that will read from standard input the rate of speed, a start time in seconds, a finish time in seconds, and an interval value in seconds. Your program should print a table containing columns for time elapsed, speed, and distance traveled, where time elapsed ranges from start time to finish time on the interval that the user entered.

9. In problem 16 of Chapter 2 you designed a solution to a problem involving half-lives of radioisotopes. Using a variation of this design, write a program that will allow the user to enter the initial weight of a certain isotope and its half-life. Your program should print a table

showing the isotope reduction in half-lives ranging from its starting weight until it reaches a weight that is less than 0.005.

10. The value of sin x is approximated by the formula

$$\sin x = x - \frac{x^3}{3!} + \frac{x^5}{5!} + \frac{x^7}{7!} + \frac{x^9}{9!} \cdots$$

where x is measured in radians. Using the function `factorial()` that was developed in problem 2, write a function with the following prototype to implement this approximation:

```
float mysin(float angle, int terms);
```

where `angle` is an angle measure in radians and `terms` is the number of terms to be used to perform this calculation. (The greater the value of `terms`, the more accurate the approximation.)

Hint: A call to `pow(-1, exp++)`, where `exp` is initialized to 1, if placed in a loop will generate the values $-1, +1, -1, +1$, alternatively. You can use this loop to help alternate the operators in your calculation loop.

11. Write a program using the function `mysin()` and the `math.h` function `sin()` to print a table comparing the values returned by both functions using a term value of 6. Your table should show the results of `mysin()` and `sin()` using angles between 0.10 and 6.0 radians on intervals of 0.1.

12. Modify your solution to problem 14 in Chapter 5 to read a series of cars from standard input and count the number of cars in each type.

13. Modify your solution to problem 15 in Chapter 5 to tally the number of accepted and rejected parts in a series of scores.

14. The ANSI *C* standard libarary (stdlib) contains a random-number generatory whose prototype is

```
int rand(void);
```

This function returns a pseudo-random integer in the range `0..RAND_MAX`, where `RAND_MAX` is at least 32,767. Write a program to generate a random number between 0 and 9 using the modulo operator. Modify the program to count the number of occurrences of each value for 1000 cycles. A true random-number generator should result in a count of 100 for each value from 0 to 9 (that is, no bias in the distribution). Is the random-number generator truly random?

15. About 1 in 90 births in the U.S. result in twins, and 1 in 3 sets of twins is identical. For 100,000 expectant parents selected at random, write a program to simulate the expected number of sets of identical twins from this group.

    Hint: Use the random-number generator and a loop from 0 to 99,999 to establish the number of sets of twins and a set of `if` statements to determine if a set of twins is identical. If a couple's odds of twins "improve" to 1 in 30 after having 1 set of twins, modify your program to consider 50,000 first-time parents and 50,000 parents of twins. Now what is the number of identical twins?

16. Many states have a lottery system in which a sequence of 6 numbers from 1 to MAX_NUMBER are chosen, where MAX_NUMBER is usually 50 to 60. Write a program using the random-number function `rand()` to generate a series of 6 numbers. Remember, once you pick a number, you can't pick it again.

7 Composite Data Types

 Scientists and engineers use computer simulations to study such things as weather patterns, earthquakes, genetic evolution of biological organisms, and the effectiveness of mechanical designs. For example, in environmental engineering computers can simulate a toxic spill and, by manipulating the variables, can establish its magnitude. The simulation can help establish how much of the soil has been permeated so that only the contaminated soil is cleaned. Factors such as soil makeup and permeability, length of time since the spill, and amount of toxic substance spilled are all taken into consideration in an effort to determine the affected area.

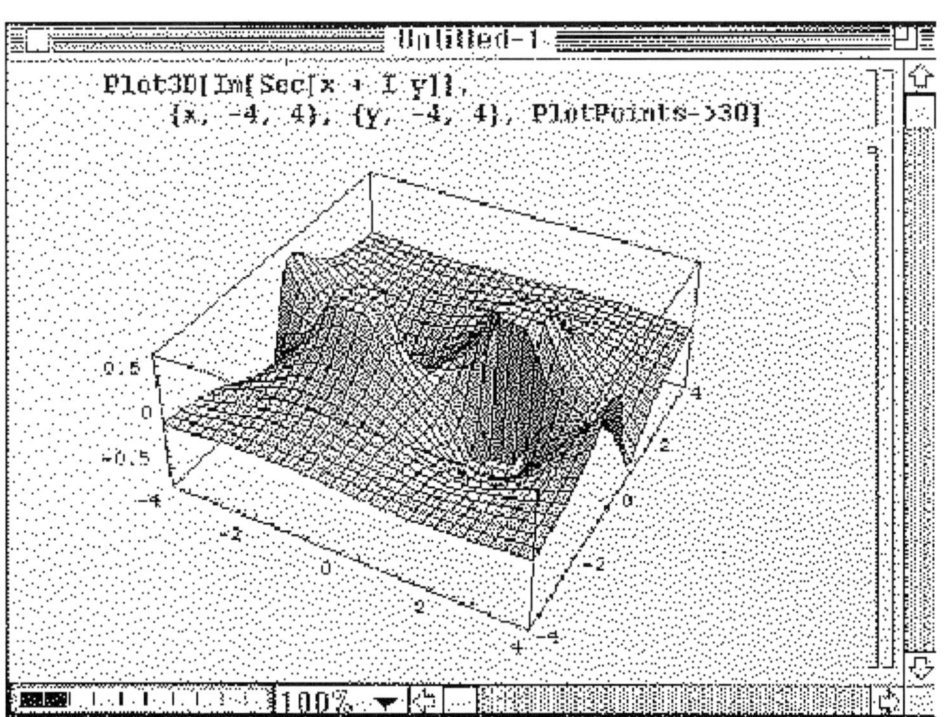

INTRODUCTION

In Chapter 2 you learned about scalar and composite data types. Until now you have been using scalar data in all of your programs. Recall that composite data are collections of related and meaningful data items, such as an address. Every *C* program has two conceptual parts: data and algorithmic logic. At the center of almost every program is the data required to solve the problem. Algorithmic logic exists to manage and manipulate the program data. To provide you with greater programming power, *C* gives you the ability to build composite data types to better represent certain kinds of information. In this chapter you will explore the most common of composite data types, the array. You will also be introduced to another composite data type in *C*, the structure.

7-1 ONE-DIMENSIONAL ARRAYS IN C

An *array* is a collection of multiple data items of the same data type. Each array item can be of only one data type, and all items in an array must be of the same type. Furthermore, an array should contain data items that have the same meaning. For instance, it would make sense to store the times to failure of a random sampling of Intel 80486™ microchips in a single array. It would *not* make sense to store employee salaries in the same array as the times to failure, because they have nothing to do with each other.

Recall that you need to declare, initialize, and manipulate program variables. Likewise, array elements undergo the same phases. Not only do arrays offer a convenient storage mechanism for data collections, but they also offer an efficient means of retrieving and manipulating items in the collection.

An array has five important components: a name, storage cells, a size, elements, and indices. The *array name* is like a variable name except that by itself, the name refers to the entire collection of data rather than a single data item. The array's *storage cells* are the memory locations that are allocated for each data item. The *array size* specifies the number of storage cells. An *array element* is a data item that occupies one of the storage cells. An *array index* is a unique numeric identifier for each storage location.

The first storage cell in an array has an index of 0, the next cell's index is 1, and so on for each cell in the array. Figure 7-1 shows how you might think of an array and all its components. This figure shows the pH factors of a collection of six rain samples taken from several sites during a thunderstorm. The pH value 7 denotes neutrality, values less than 7 denote acidity, and values greater than 7 denote alkalinity.

The array in Figure 7-1 is named `acid_rain_data`, and each array element represents the pH factor of a single sample. This array is a one-dimensional array. A *one-dimensional array* is an array that is visualized as a single row of elements in sequence. Although the array size in Figure 7-1 is 8, only six elements are currently stored in the array. Cells 6 and 7 remain empty.

You use subscript notation to refer to a single array element. *Subscript notation* is the combination of an array name and an array index value that references a single element in the array. For example, if you wish to print the contents of storage cell 3, you make the following call to `printf()`:

```
printf("%f\n", acid_rain_data[3]);          /* 9.1 is printed */
```

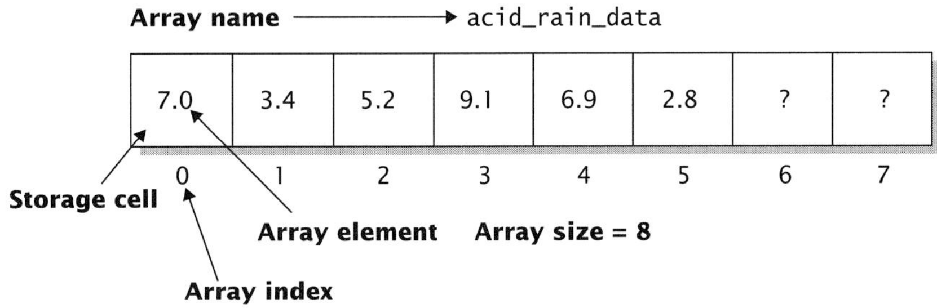

Figure 7-1 Visualizing an Array

The [3] in this statement is the *array subscript.* You will learn how to use array subscripts inside `for` loops to process each element in an array.

In the following sections you will learn how to declare and initialize an array as well as how to process the elements in an array.

Declaring an Array

You declare an array by specifying the array type, name, and size. The subscript notation specifies the array's size. For example, you declare the `acid_rain_data` array as follows:

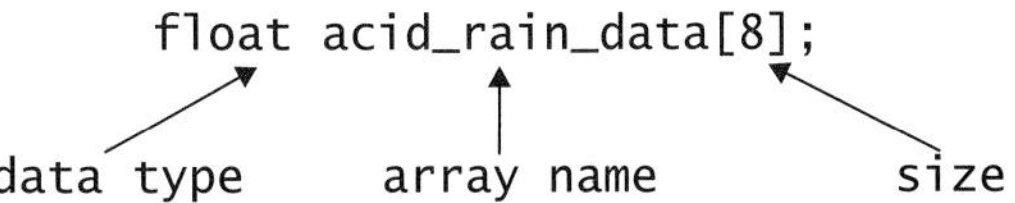

This declaration tells the compiler to allocate memory storage for 8 data items of type `float` and group them into a collection called `acid_rain_data`. The data type in the array declaration describes the type of each array element.

Note that the subscript 8 in the declaration does not refer to storage cell number 8. The 8 describes the maximum number of elements in the array. Because the array indices begin at 0, there is no cell number 8, even though there are a total of 8 storage cells (0–7) in the array. Keep this numbering in mind as we discuss arrays, because it is occasionally a point of confusion for new *C* programmers.

The general form of a one-dimensional array declaration is

```
<data type> <array name> '['<size>']';
```

Where `<data type>` may be any data type, `<array name>` may be any legal identifier name, and `<size>` may be any integer value as long as there is enough memory to allocate the requested space. In this general form, the square brackets are surrounded in single quotes because they are to be taken literally rather than to denote optional inclusion.

Although an array can be declared of any data type, you cannot mix the data types of the storage cells in the same array. The rule to remember is that the items in an array must be of the same type and must have the same meaning or purpose. The size of your array is limited only by the computer's memory availability.

Try It  Write a declaration for arrays that are to contain

- Twenty-five samples of a stream temperature
- Fifty student scores for a midterm exam
- The distances from the Earth of 1000 stars in the Andromeda galaxy as measured in light years
- Flow rate, in cubic feet per second, of the Colorado River flowing from the Glen Canyon hydroelectric dam at exactly 5:00 p.m. on 30 consecutive days

---

Initializing an Array

As with scalar variables, an array contains nothing meaningful when it is declared; therefore, you need to initialize an array using it. The simplest way to initialize small arrays is in the declaration. Following the declaration, you use the simple assignment operator and merely list the contents of each storage cell. Each value in the initialization list is separated by a comma, and the entire list is surrounded by curly brackets. You can initialize the storage cells of `acid_rain_data` to contain 0.0 with the following declaration:

```
float acid_rain_data[8] = {0.0, 0.0, 0.0, 0.0, 0.0, 0.0, 0.0, 0.0};
```

Using this technique, the first value in the initialization list is stored in the first storage cell of the array, the second value in the second cell, and so on. If there are fewer values in the list than storage cells, the remaining cells will contain meaningless values. If there are more values in the list than storage cells, the compiler will issue an error message. Although we initialized `acid_rain_data` to contain 0.0 in each cell, any list of eight real numbers is acceptable.

Although this method of initialization is simple for small arrays, it becomes tedious for large arrays. The `for` loop offers an elegant and efficient means of initializing arrays of any size. Because the technique for accessing the storage cells in an array involves numeric indices, you can use the loop-control variable as the subscript to loop through each cell in the array. The program in Example 7-1 demonstrates this use of the `for` loop.

EXAMPLE 7-1   ## Using a `for` Loop to Initialize an Array

```
/*-----------------------------------------------------------------
   Description: This program will initialize an array to contain
      0.0 in each cell.
   Programmer: Ken Collier
   Date of Last Revision: 1-1-1995
   Modifications
        Date            Description
        none            none

-------------------------------------------------------------------*/
```

```c
#include <stdio.h>

#define ARRAY_SIZE 8

void main(void)
{
  float acid_rain_data[ARRAY_SIZE];
  int lcv;                /* Loop Control Variable */

  /* Initialize array */
  for(lcv=0; lcv < ARRAY_SIZE; ++lcv)
    acid_rain_data[lcv] = 0.0;

  /* Print out the contents of the array */
  for(lcv=0; lcv < ARRAY_SIZE; ++lcv)
    printf("Element %d contains %f\n", lcv, acid_rain_data[lcv]);
}
```

. . . .

Let's step through the loop in boldface type in Example 7-1. At loop entry, lcv is initialized to 0, and the loop will iterate as long as lcv is less than 8. Therefore, on the first execution of the loop body, acid_rain_data[0] is set to 0.0. On the next iteration, lcv is incremented to 1, and acid_rain_data[1] is set to 0.0. Using iterative looping, two lines of code will initialize each cell in the array.

The output of the program is

```
Element 0 contains 0.000000
Element 1 contains 0.000000
Element 2 contains 0.000000
Element 3 contains 0.000000
Element 4 contains 0.000000
Element 5 contains 0.000000
Element 6 contains 0.000000
Element 7 contains 0.000000
```

Several concepts are demonstrated by this program. First, you can use a variable as an array subscript instead of a constant. If it weren't for this feature, you would have to write eight separate statements to initialize each storage cell to 0.0. Second, the use of symbolic constants to define array sizes adds flexibility to your programs. Notice that the symbolic constant ARRAY_SIZE in Example 7-1 is used in the array declaration as well as in both for loop headers. Suppose you wanted to measure 50 rain samples. You simply change the preprocessor statement to

```
#define ARRAY_SIZE 50
```

Because the rest of the program uses the symbolic constant rather than the literal constant 8, this simple change will update all references during compilation. This use of symbolic constants to define array sizes is a good programming habit to adopt.

Another characteristic of arrays is that each individual array element can be manipulated exactly as if it were a scalar variable. In the previous program, the statement

```
acid_rain_data[lcv] = 0.0;
```

contains a subscripted array reference as the left operand of an assignment operation. You can also use a subscripted array reference as the operand in an arithmetic expression, the argument to a function, or anyplace that a simple variable can be used.

Although array elements are most commonly initialized to contain the same value, this is not always the case. Suppose you want to write a program in which an array of 100 elements is initialized so that the odd-indexed cells contain the value 5 and the even-indexed cells contain the value 10. There are several ways to write such a program. One approach would be to write two separate loops: one loop to visit each of the cells with an even index and the other to visit each cell with an odd index. Example 7-2 contains a program that implements this idea. The loop-control variable in the first loop is initialized to 0 and incremented by 2 on each iteration so that it visits each even index. The loop-control variable in the second loop is initialized to 1 and incremented by 2 on each iteration so that it visits each odd index.

EXAMPLE 7-2 ██ **Initializing Even Cells to 10 and Odd Cells to 5**

```
/*----------------------------------------------------------------
   Description: This program will initialize the even-indexed
       cells of an array to 10 and the odd-indexed cells to 5
       using two separate for loops.
   Programmer: Ken Collier
   Date of Last Revision: 1-1-1995
   Modifications
           Date            Description
           none            none
-----------------------------------------------------------------*/

#include <stdio.h>

#define ARRAY_SIZE 100

void main(void)
{
  int my_array[ARRAY_SIZE],
      lcv;                 /* Loop Control Variable */

  /* First visit all of the even indexed cells */
  for(lcv=0; lcv < ARRAY_SIZE; lcv += 2)
    my_array[lcv] = 10;
```

```
/* Now visit all of the odd indexed cells */
for(lcv=1; lcv < ARRAY_SIZE; lcv += 2)
  my_array[lcv] = 5;

for(lcv=0; lcv < ARRAY_SIZE; ++lcv)
  printf("%d\n", my_array[lcv]);
}
```

. . . .

The first ten lines of the output of this program are

```
10
5
10
5
10
5
10
5
10
5
```

Try It   ◆ Write a simple program that declares an array of 50 integers and initializes each cell in the array to contain the value 100.
◆ Modify your program so that the array contains 500 elements

Processing Array Elements

As you saw in the previous section, the for loop is an efficient way to initialize each cell in an array. More generally, a for loop is an efficient way to perform any set of actions on each element in an array.

We will use the acid rain example as the basis for several examples of a for loop to process array elements. Environmental engineers are concerned with the effects of high acid or alkaline levels in rain and their effect on the environment. Acid rain is a result of anthropogenic (human-generated) gaseous emissions (specifically, sulfur dioxide and nitrous oxide) that combine with water vapor in the atmosphere to form sulfuric and nitric acids. Unpolluted rain is naturally acidic due to the presence of CO^2 and the formation of carbonic acid. The natural pH of pure rain is about 5.6. Average pH values over the northeastern United States and southeastern Canada have been recorded at values of 4.2 to 4.5.

Sulfuric acid in the atmosphere separates into hydronium ions and sulfates. The hydronium ions that fall to Earth along with rainfall cause an increase in the rain's acidity.

The formula to calculate pH is defined as the negative $\log_{10}$ of the molar concentration of the hydrogen ions [H+] that occurs in the disassociation of compounds in solution, or

$$pH = -\log_{10}\left(\frac{\text{moles of hydronium ions}}{\text{liter}}\right)$$

It is impossible to have pH values less than 0 because a solution composed solely of hydrogen ions would have a molar concentration of 1.0 and thus a pH of 0.

Suppose you are to write a program that reads in the concentration of hydronium ions in a 1-liter water sample, then calculates and prints its pH factor. The algorithm for this program might be

```
Declare variable to store rain sample
Initialize input data variable to 0.0
Read in rainwater data from standard input
Convert rainwater data to pH factor
Print the pH factor
```

Example 7-3 contains a program based on this algorithm.

EXAMPLE 7-3 Calculating the pH of a Rainwater Sample

```
/*-----------------------------------------------------------------
   Description: This program calculates and prints the
       pH of a rainwater sample. pH factors
       are calculated using the relationship

          pH = -log10(moles of hydronium ions/liter)

   Programmer: Ken Collier
   Date of Last Revision: 1-1-1995
   Modifications
          Date              Description
          none              none
   ------------------------------------------------------------------*/

#include <stdio.h>
#include <math.h>

void main(void)
{
   float acid_rain_data;/* Hydronium Ion Concentration */

   /* Initialize Variable */
   acid_rain_data = 0.0;

   /* Read rainwater data from standard input */
   printf("Enter data for sample: ");
   scanf("%f", &acid_rain_data);

   /* Convert rainwater data to pH factor */
   acid_rain_data = -log10(acid_rain_data);

   printf("The sample has a pH of %f.\n", acid_rain_data);
}
```

· ·

Given a hydronium concentration of `0.00000005`, the output of this program is

```
Enter data for sample: .00000005
The sample has a pH of 7.301030.
```

Now, instead of performing this calculation on a single water sample, let's write a program to store the hydronium concentration of eight water samples in an array and calculate and print the average pH. This change requires the following algorithm:

```
Declare array to store rain sample data
Initialize each array element to 0.0
Read rain data into array
Convert rain data to pH factor
Calculate average pH
Print average pH
```

Notice that this algorithm is almost identical to the previous algorithm. The primary difference is that each step is performed on each array element rather than on a single data item. You do this by using a for loop to perform each action on each array element. Additionally, we must calculate the average pH. In previous chapters you have seen simple programs that calculate averages using a loop to sum the numbers and then dividing by the number of numbers. The program in Example 7-4 uses the same approach. In this example, the first for loop initializes the array, the second is used to read rainwater data, the third converts hydronium concentrations to the pH scale, and the fourth calculates the average pH. Aside from the for loops surrounding each action or set of actions, the program in Example 7-4 is very similar to the program in Example 7-3.

EXAMPLE 7-4   Calculating the Average pH of Rain Samples

```
/*---------------------------------------------------------------
   Description: This program will calculate and print the average
       pH of several rainwater samples. pH factors
       are calculated using the relationship

           pH = -log10(moles of hydronium ions/liter)

   Programmer: Ken Collier
   Date of Last Revision: 1-1-1995
   Modifications
        Date               Description
        none               none
   ------------------------------------------------------------*/

#include <stdio.h>
#include <math.h>

#define SAMPLE_SIZE 8
```

```c
void main(void)
{
  float   acid_rain_data[SAMPLE_SIZE],
                                     /* Hydronium concentrations */
          total = 0.0,               /* Sum of pH factors */
          average;                   /* Average acidity */
  int     lcv;                       /* Loop control variable */

  /* Initialize array */
  for (lcv=0; lcv < SAMPLE_SIZE; ++lcv)
    acid_rain_data[lcv] = 0.0;

  /* Read rainwater data from standard in */
  for (lcv=0; lcv < SAMPLE_SIZE; ++lcv){
    printf("Enter data for sample %d: ", lcv+1);
    scanf("%f", &acid_rain_data[lcv]);
  }

  /* Convert rainwater data to pH factor */
  for (lcv=0; lcv < SAMPLE_SIZE; ++lcv)
    acid_rain_data[lcv] = -log10(acid_rain_data[lcv]);

  /* Calculate average acidity */
  for (lcv=0; lcv < SAMPLE_SIZE; ++lcv)
    total += acid_rain_data[lcv];
  average = total / SAMPLE_SIZE;

  printf("The average pH is %f.\n", average);
}
```

. . . .

Given the input values 0.00023, 0.0004, 0.00005, 0.001, 0.0003, 0.00022, 0.0009, and 0.0004, the output of this program is

```
Enter data for sample 1: .00023
Enter data for sample 2: .0004
Enter data for sample 3: .00005
Enter data for sample 4: .001
Enter data for sample 5: .0003
Enter data for sample 6: .00022
Enter data for sample 7: .0009
Enter data for sample 8: .0004
The average pH is 3.495174.
```

In this example you used the symbolic constant SAMPLE_SIZE to determine the upper bound of the for loop. You could have prompted the user to enter the number of samples to be input and used this value as the upper bound. For this technique to work, the value entered by the user must be less than or equal to the size of the array. If the upper bound is less than the array size, some array elements will be ignored.

The standard technique for processing the elements in an array is the use of for loops. Anytime you wish to visit each cell in an array, whether it is to retrieve or modify the contents, you should use a for loop.

What If  In the program of Example 7-4, the reading of hydronium ion concentration data and the conversion of that data to the pH scale are performed in two separate for loops. It would be more efficient to perform these actions in the same for loop. Modify this program so that these two steps are performed in the same for loop.

7-2 PASSING ARRAYS AS PARAMETERS

An array name with no subscript is simply a reference to the address of the first array element. Like other data items, arrays can be passed as parameters to *C* functions. Of course, the function receiving an array must be defined to accept an array as a parameter. Passing arrays into functions is achieved by declaring the parameter to be an array. Example 7-5 shows a function to read data into the acid_rain_data array in Example 7-4.

EXAMPLE 7-5 ### Arrays as Function Parameters

```
void load_array(float data[SAMPLE_SIZE])
{
  int lcv;             /* Loop control variable */

  /* Read rainwater data from standard in */
  for (lcv=0; lcv < SAMPLE_SIZE; ++lcv){
    printf("Enter data for sample %d: ", lcv+1);
    scanf("%f", &data[lcv]);
  }
}
```

. . . .

Notice that the body of this function is nearly identical to the code segment in Example 7-4 that is responsible for reading data into the array. The only exception is that the array being processed in the for loop is referred to by its formal parameter name data rather than its actual argument name acid_rain_data.

The name of an array by itself (with no subscript) is a reference to the entire array. Therefore, a call to the function defined in Example 7-5 would look like

```
load_array(acid_rain_data);
```

The reference to acid_rain_data is actually a reference to the address of the first memory location of the acid_rain_data array. Therefore, arrays are passed as parameters by reference rather than by value, and there is no need to use the address-of operator when passing arrays as parameters. When an array name is passed as an argument to a function, the formal parameter becomes a reference to the entire array, and subscript notation can be used to access its elements. All changes made to an array through this formal parameter affect the actual array. Memory is saved when the formal parameter is a reference to the actual array memory because a copy of the entire array is not necessary.

Try It   Write a function named `calculate_pH_factors()` that will accept the `acid_rain_data` array as an actual argument and will convert each element from raw data to a pH value.

7-3   TWO-DIMENSIONAL ARRAYS IN C

One-dimensional arrays are useful when your data collection can be represented as a simple sequence of data items. But sometimes the relationship of the data items to one another is as important as the data itself. For example, suppose your acid-rain program must handle rainwater samples collected from eight different sites over five consecutive rainfalls. This type of data is best stored in a table of five rows by eight columns. In this table, each row represents a different rainfall, and each column represents a different sampling site. Figure 7-2 shows a table in this format. In a program, a table such as this is called a *two-dimensional array.*

A two-dimensional array supports the idea that subgroupings of data items are as important as the data values. If you place all 40 hydronium-concentration values into a one-dimensional array, you lose important sample sites and rainfall information. You should always use a two-dimensional array if there are two attributes by which your data can be grouped. In the case of our acid-rain example, these attributes are (1) from which rainfall the sample was taken and (2) from which site the sample was taken.

Declaring an Array

A two-dimensional array is declared very much like a one-dimensional array except that you must specify the size of both dimensions following the array name. To declare the array in Figure 7-2, you write

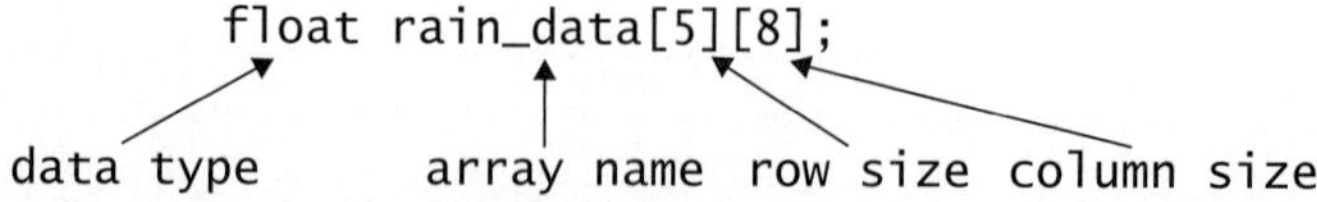

Note that the number of rows appears first and the number of columns appears second. This ordering is called *row-major order.*

The row and column index values begin with 0, which is shown along the bottom and right edges of the array in Figure 7-2. As with one-dimensional arrays, you use subscript notation to access individual array elements. You use row-major order in this subscript notation just as you did in the decla-

	Site A	Site B	Site C	Site D	Site E	Site F	Site G	Site H	
Rainfall 1	0.00025	0.00004	0.0003	0.00174	0.0008	0.00001	0.0009	0.00045	0
Rainfall 2	0.0003	0.00003	0.00022	0.002	0.00055	0.0001	0.0007	0.0004	1
Rainfall 3	0.0002	0.000042	0.00035	0.0015	0.00058	0.00008	0.00085	0.00042	2
Rainfall 4	0.00028	0.000035	0.00032	0.00221	0.00083	0.00005	0.0013	0.00048	3
Rainfall 5	0.00022	0.000045	0.00039	0.0018	0.00055	0.00009	0.00088	0.0005	4
	0	1	2	3	4	5	6	7	

Figure 7-2 Two-Dimensional Storage of Rainwater Data

ration. So, to print the hydronium concentration of the sample taken at site D during rainfall 3, you use the *C* statement

```
printf("Site D, Rainfall 3: %f\n", rain_data[2][3]);
```

In general, the form of a two-dimensional array declaration is

```
<data type> <array name> '['<rowsize>']' '['<columnsize>']';
```

The total size of the array can be calculated by multiplying the row dimension by the column dimension.

Initializing an Array

A two-dimensional array can be initialized in the declaration just like a one-dimensional array; however, in addition to grouping the complete set of values in a pair of curly brackets, each row of values must also be enclosed in curly brackets. To declare the `rain_data` array and initialize its cells to contain 0.0, you write

```
float rain_data[5][8] = {{0.0, 0.0, 0.0, 0.0, 0.0, 0.0, 0.0, 0.0},
                         {0.0, 0.0, 0.0, 0.0, 0.0, 0.0, 0.0, 0.0},
                         {0.0, 0.0, 0.0, 0.0, 0.0, 0.0, 0.0, 0.0},
                         {0.0, 0.0, 0.0, 0.0, 0.0, 0.0, 0.0, 0.0},
                         {0.0, 0.0, 0.0, 0.0, 0.0, 0.0, 0.0, 0.0}};
```

This approach clearly becomes cumbersome for all but the smallest arrays. Again, the `for` loop is a more general-purpose method for array initialization. To initialize a two-dimensional array, you must use two nested `for` loops, one for each dimension. The strategy is to visit each cell in the first row, then each cell in the second row, and so on until all cells have been visited.

EXAMPLE 7-6   Initializing a Two-Dimensional Array Using for Loops

```
/*-----------------------------------------------------------------
     Description: This program will initialize a two-dimensional
          array to contain 0.0 in each cell.
     Programmer: Ken Collier
     Date of Last Revision: 1-1-1995
     Modifications
               Date              Description
               none              none
     ---------------------------------------------------------------*/

#include <stdio.h>

#define MAX_ROWS 5
#define MAX_COLS 8

void main(void)
{
  float rain_data[MAX_ROWS][MAX_COLS];
  int row, /* Loop control variable for row loop */
      col; /* Loop Control Variable for column loop */

  /* Initialize array */
  for(row=0; row < MAX_ROWS; ++row)
    for(col=0; col < MAX_COLS; ++col)
      rain_data[row][col] = 0.0;

  /* Print out the contents of the array */
  for(row=0; row < MAX_ROWS; ++row){
    for(col=0; col < MAX_COLS; ++col)
      printf("%f ", rain_data[row][col]);
    putchar('\n');
  }
}
```

. . . .

Although not very interesting, the output of this program shows you that the array cells were successfully initialized to 0.0.

```
0.00 0.00 0.00 0.00 0.00 0.00 0.00 0.00
0.00 0.00 0.00 0.00 0.00 0.00 0.00 0.00
0.00 0.00 0.00 0.00 0.00 0.00 0.00 0.00
0.00 0.00 0.00 0.00 0.00 0.00 0.00 0.00
0.00 0.00 0.00 0.00 0.00 0.00 0.00 0.00
```

Let's trace the initialization loops so that you can understand what is happening. First the outer loop control variable row is set to 0. Next the inner loop-control variable col is set to 0. Then the body of the inner loop is executed, and rain_data[0][0] is set to 0.0. Because the body of the outer loop is the entire inner loop, the inner loop will iterate eight times for each iteration of the outer loop. Therefore, the value 0.0 will be assigned to `rain_data[0][0]`, `rain_data[0][1]`, `rain_data[0][2]`, ... , `rain_data[0][7]`. On the second iteration of the outer loop, `row` will be incremented to 1 and each of the cells in the next row will be visited. The loops continue executing until `row` is set incremented to 5 and `col` is incremented to 8, at which point all cells in the array have been initialized. As with the processing of one-dimensional arrays, the loop-control variables in the `for` loops are used as the subscript values for accessing array elements.

Notice again the use of symbolic constants to define the array dimensions. As with one-dimensional arrays, this use of symbolic constants helps you write programs that are easy to modify. Because arrays are almost always initialized and processed using `for` loops, you should always define symbolic constants to specify array sizes. You should use either these symbolic constants or user-input values as the upper boundaries for array-processing loops.

Processing Array Elements

As with initialization, the elements of a two-dimensional array are best processed using a set of nested `for` loops. Extending the acid-rain example, suppose in addition to the overall average pH you wish to calculate the average at each collection site for all five rainfalls as well as the average of all collection sites per rainfall. The following algorithm describes the high-level steps necessary to solve this problem:

```
Declare array to store rain sample data
Initialize array elements to 0.0
Read rain data
Convert rain data to pH scale
Calculate overall average pH
Calculate average pH for each site
Calculate average pH for each rainfall
```

Visualizing the rain-data array as in Figure 7-2, consider what must happen to solve each step in the algorithm. We solved the first step of initializing the array in the previous section.

As mentioned earlier, the reading of rain data and converting of the data to a pH scale can be accomplished in the same looping construct. This process is best described in the following algorithm:

```
for each row in rain_data
  for each column in the current row of rain_data
    read data item into rain_data[row][column]
    convert value in rain_data[row][column] to pH scale
```

Similarly, to calculate the overall average, we will visit each cell of the array and add the value of the cell into a running total. After all cells have been visited, we will divide the running total by the number of cells. This

	Site A	Site B	Site C	Site D	Site E	Site F	Site G	Site H
Rainfall 1	3.60	4.40	3.52	2.76	3.10	5.00	3.05	3.35
Rainfall 2	3.52	4.52	3.66	2.70	3.26	4.00	3.15	3.40
Rainfall 3	3.70	4.38	3.46	2.82	3.24	4.10	3.07	3.38
Rainfall 4	3.55	4.46	3.49	2.66	3.08	4.30	2.89	3.32
Rainfall 5	3.66	4.35	3.41	2.74	3.26	4.05	3.06	3.30

(Column indices: 0 1 2 3 4 5 6 7)

rainfall_averages:

index	value
0	3.60
1	3.53
2	3.52
3	3.47
4	3.48

site_averages:

3.61	4.42	3.51	2.74	3.19	4.29	3.04	3.35
0	1	2	3	4	5	6	7

Figure 7-3 Relationships Among Three Arrays

strategy is a generalization of the average calculation in Example 7-4, and the following algorithm reflects this approach:

```
total = 0.0
for each row in rain_data
   for each column in the current row of rain_data
      add rain_data[row][column] into total
average = total / (MAX_ROWS*MAX_COLS)
```

Now, to calculate the average of each site and the average of each rainfall, you need to consider that each row in the array represents a separate rainfall and each column represents the data collected at each site. Therefore, to calculate the average pH per rainfall, you must calculate the average of each row separately. To calculate the average pH per site, you must calculate the average of each column separately. Because we need some means of storing each of these averages, it makes sense to create two additional one-dimensional arrays, one to store the averages of each rainfall and the other to store the averages of each site. We will call these new arrays `rainfall_averages` and `site_averages`, respectively. Figure 7-3 will help you visualize the relationships among the `rain_data` array and the new arrays. The average of the first row in `rain_data` is stored as the first element in `rainfall_averages`, the average of the second row as the second element, and so on. Similarly, the average of the first column of `rain_data` is stored as the first element of `site_averages`, the average of the second column as the second element, and so on.

The algorithm for calculating the average of each row is

```
for each row in rain_data
  reset total to 0.0
  for each column in the current row
    total = total + rain_data[row][column]
  rainfall_averages[row] = total / MAX_COLS
```

Notice that two for loops are still needed to visit each array element. However, just prior to processing a new row, total must be reinitialized to 0.0. Placing the assignment statement

```
total = 0.0
```

just before the inner for loop causes this reinitialization to occur. Also, the average for a given row must be calculated immediately after that row has been processed and before the total is reinitialized. Placing the statement

```
rainfall_averages[row] = total / MAX_COLS
```

just after the inner for loop calculates the row average. The calculation and storage of site averages can be accomplished by the use of a similar algorithm.

```
for each column in rain_data
  reset total to 0.0
  for each row in the current column
    total = total + rain_data[row][column]
  site_averages[column] = total / MAX_ROWS
```

This algorithm is almost identical to the previous algorithm except that it processes the two-dimensional array in a column-by-column order rather than a row-by-row order.

Example 7-7 shows the combination of these algorithms into a program that will convert all data to a pH scale, calculate the overall pH average, calculate the average pH of all sites per rainfall, calculate the average pH of all rainfalls per site, and then print all averages.

EXAMPLE 7-7   ## Calculating pH Statistics

```
/*------------------------------------------------------------------
  Description: This program calculates and prints:
    - The overall average pH of 8 rainfall samples
      taken over 5 rainfalls.
    - The average pH per rainfall of all sites.
    - The average pH per site over 5 rainfalls.

    pH factors are calculated using the relationship

      pH = -log10(moles of hydronium ions/liter)
```

```
      Programmer: Ken Collier
      Date of Last Revision: 1-1-1995
      Modifications
            Date            Description
            none            none
-----------------------------------------------------------------*/

#include <stdio.h>
#include <math.h>

/* Symbolic Constants */

#define MAX_SITES 8/* Number of columns in array */
#define MAX_RAINFALLS 5/* Number of rows in array */

/* Function Prototypes */

void init_array(float data[MAX_RAINFALLS][MAX_SITES]);
void read_and_convert_rain_data(float data[MAX_RAINFALLS][MAX_SITES]);
float calc_overall_average(float data[MAX_RAINFALLS][MAX_SITES]);
void calc_average_per_rainfall(float data[MAX_RAINFALLS][MAX_SITES],
                  float avgs[MAX_RAINFALLS]);
void calc_average_per_site(float data[MAX_RAINFALLS][MAX_SITES],
                  float avgs[MAX_SITES]);
void output_site_averages(float data[MAX_SITES]);
void output_rainfall_averages(float data[MAX_RAINFALLS]);

void main(void)
{
  floatrain_data[MAX_RAINFALLS][MAX_SITES],
    site_averages[MAX_SITES],
              /* Stores separate site stats */
    rainfall_averages[MAX_RAINFALLS],
              /* Stores separate rainfall stats */
    overall_average;/* Average pH */
  int lcv;      /* Loop control variable */

  /* Initialize all arrays */
  init_array(rain_data);

  for (lcv=0; lcv < MAX_RAINFALLS; ++lcv)
    rainfall_averages[lcv] = 0.0;

  for (lcv=0; lcv < MAX_SITES; ++lcv)
    site_averages[lcv] = 0.0;

  read_and_convert_rain_data(rain_data);

  /* Calculate statistics */
  overall_average = calc_overall_average(rain_data);
  calc_average_per_rainfall(rain_data, rainfall_averages);
  calc_average_per_site(rain_data, site_averages);
```

```c
    /* Output overall average, site averages, & rainfall averages */
    printf("\nThe overall average pH is %.2fpH.\n\n",
            overall_average);
  output_site_averages(site_averages);
  output_rainfall_averages(rainfall_averages);
}


/*-----------------------------------------------------------------
Module: init_array()
Description: Initializes a 2D array.
Input Parms: A 2D array of MAX_RAINFALLS x MAX_SITES
Returns: None
Modifications
        Date            Description
        none            none
-----------------------------------------------------------------*/
void init_array(float data[MAX_RAINFALLS][MAX_SITES])
{
  int row, col;/* Loop control variables */

  /* Initialize rain data array */
  for (row=0; row < MAX_RAINFALLS; ++row)
    for (col=0; col < MAX_SITES; ++col)
      data[row][col] = 0.0;
}


/*-----------------------------------------------------------------
Module: read_and_convert_rain_data()
Description: Loads a 2D array with rain data from standard input
  and converts it to pH factor.
Input Parms: A 2D array of MAX_RAINFALLS x MAX_SITES
Returns: None
Modifications
        Date            Description
        none            none
-----------------------------------------------------------------*/
void read_and_convert_rain_data(float data[MAX_RAINFALLS][MAX_SITES])
{
  int row, col;/* Loop control variables */
  char site;/* For printing site labels */

  /* Read rainwater data from standard in & convert to pH scale */
  for (row=0; row < MAX_RAINFALLS; ++row){
    site = 'A';
    for (col=0; col < MAX_SITES; ++col){
      printf("Enter data for site %c during rainfall %d: ",
            site++, row+1);
      scanf("%f", &data[row][col]);
      data[row][col] = -log10(data[row][col]);
    }
  }
}
```

```c
/*-----------------------------------------------------------------
Module: calc_overall_average()
Description: Averages the pH factors in a 2D array.
Input Parms: A 2D array of MAX_RAINFALLS x MAX_SITES
Returns: None
Modifications
        Date               Description
        none               none
------------------------------------------------------------------*/
float calc_overall_average(float data[MAX_RAINFALLS][MAX_SITES])
{
  int row, col;/* Loop control variables */
  float total=0.0;/* Running total */

  /* Calculate overall average pH */
  for (row=0; row < MAX_RAINFALLS; ++row)
    for (col=0; col < MAX_SITES; ++col)
      total += data[row][col];

  return((float)total / ((float)MAX_RAINFALLS * (float)MAX_SITES));
}


/*-----------------------------------------------------------------
Module: calc_average_per_rainfall()
Description: Averages each row in a 2D array.
Input Parms: A 2D array of MAX_RAINFALLS x MAX_SITES
Returns: None
Modifications
        Date               Description
        none               none
------------------------------------------------------------------*/
void calc_average_per_rainfall(float data[MAX_RAINFALLS][MAX_SITES],
                  float avgs[MAX_RAINFALLS])
{
  int row, col;/* Loop control variables */
  float total=0.0;/* Running total */

  /* Calculate the average of each row */
  for (row=0; row < MAX_RAINFALLS; ++row){
    total = 0.0;
    for (col=0; col < MAX_SITES; ++col)
      total += data[row][col];
    avgs[row] = total / MAX_SITES;
  }
}


/*-----------------------------------------------------------------
Module: calc_average_per_site()
Description: Averages each column in a 2D array.
Input Parms: A 2D array of MAX_RAINFALLS x MAX_SITES
Returns: None
```

```c
Modifications
        Date                Description
        none                none
------------------------------------------------------------------*/
void calc_average_per_site(float data[MAX_RAINFALLS][MAX_SITES],
                  float avgs[MAX_SITES])
{
  int row, col;/* Loop control variables */
  float total=0.0;/* Running total */

  /* Calculate the average of each column */
  for (col=0; col < MAX_SITES; ++col){
    total = 0.0;
    for (row=0; row < MAX_RAINFALLS; ++row)
      total += data[row][col];
    avgs[col] = total / MAX_RAINFALLS;
  }
}


/*----------------------------------------------------------------
Module: output_site_averages()
Description: Displays each element in 1D array.
Input Parms: A 1D array of MAX_SITES elements.
Returns: None
Modifications
        Date                Description
        none                none
------------------------------------------------------------------*/
void output_site_averages(float data[MAX_SITES])
{
  int lcv;/* Loop Control Variable */
  char site;/* For printing site label */

  /* Output site averages */
  site = 'A';
  for (lcv=0; lcv < MAX_SITES; ++lcv)
    printf("The average for site %c is %.2fpH.\n",
         site++, data[lcv]);
  printf("\n");
}


/*----------------------------------------------------------------
Module: output_rainfall_averages()
Description: Displays each element in 1D array.
Input Parms: A 1D array of MAX_RAINFALLS elements.
Returns: None
Modifications
        Date                Description
        none                none
------------------------------------------------------------------*/
```

```
void output_rainfall_averages(float data[MAX_RAINFALLS])
{
  int lcv;/* Loop Control Variable */
  char site;/* For printing site label */

  /* Output rainfall averages */
  for (lcv=0; lcv < MAX_RAINFALLS; ++lcv)
    printf("The average for rainfall %d is %.2fpH.\n",
           lcv+1, data[lcv]);
}
```

. . . .

Given the input values shown in Figure 7-2, the output of this program is

```
The overall average pH is 3.52pH.
The average for site A is 3.61pH.
The average for site B is 4.42pH.
The average for site C is 3.51pH.
The average for site D is 2.74pH.
The average for site E is 3.19pH.
The average for site F is 4.29pH.
The average for site G is 3.04pH.
The average for site H is 3.35pH.


The average for rainfall 1 is 3.60pH.
The average for rainfall 2 is 3.53pH.
The average for rainfall 3 is 3.52pH.
The average for rainfall 4 is 3.47pH.
The average for rainfall 5 is 3.48pH.
```

Notice the heavy use of nested for loops for the initialization, loading, and processing of array elements in this program.

What If  Change the program in Example 7-7 so that it monitors data from 10 different sites over 15 rainfalls. Run the program to be certain your change works. How difficult is it to make this change?

7-4 ARRAYS OF MORE THAN TWO DIMENSIONS

Some applications require arrays of more than two dimensions. For example, suppose you wish to write a university record-keeping program. In this program, you need to keep track of the number of students separated by gender, major, and classification. This separation requires a *three-dimensional array,* one dimension for each characteristic.

In general, you can declare arrays in *C* to be of as many dimensions as memory will allow. To calculate the amount of memory consumed by an array, you simply multiply the number of bytes required to store a single array element by the size of each dimension in the array. Because a data item of type `float` consumes 4 bytes on most computers, the amount of memory consumed by the `rain_data` array is $4 \times 5 \times 8$ bytes (160 bytes).

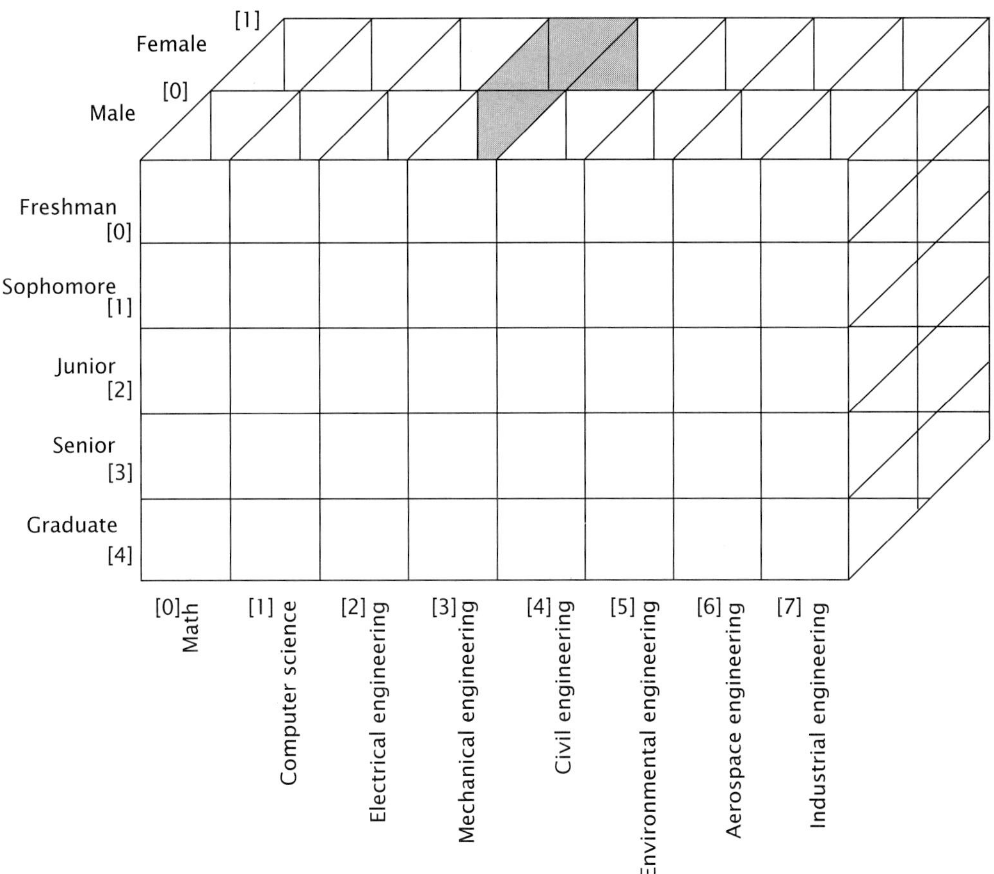

Figure 7-4 Visualizing a Three-Dimensional Array

The number of elements in the array is calculated by multiplying the size of each dimension together.

In general, the form of any array declaration is

```
<data type> <array name> '['<size>']' [<dimension list>];
```

where

```
<dimension list> ::= '['<size>']' [<dimension list>]
```

This general form description tells you that you can declare an array to be of one or more dimensions by simply listing the size of each dimension in subscript notation following the array name. The `student_population` array would be declared by the statement

```
int student_population[MAX_CLASS][MAX_MAJORS][GENDERS];
```

You can visualize a three-dimensional array as a cube. Beyond three dimensions, arrays are difficult to visualize and are seldom needed.

As with one- and two-dimensional arrays, arrays of more than two dimensions are processed using nested `for` loops. To visit each element in a multidimensional array, you need a `for` loop for each dimension. Figure 7-4 contains a graphical representation of the `student_population` array. Each cell is intended to contain the number of students with the same combination of characteristics. For example, the number of freshman female students majoring in mechanical engineering is stored in the shaded cell and is accessed by the expression

```
student_population[0][3][1]
```

Notice that the order of dimensions is rows, columns, and layers, which is an extension of the two-dimensional array order. Example 7-8 shows a program segment that will print each element of this array after it has been filled with data.

EXAMPLE 7-8   ## Initializing a Three-Dimensional Array

```
#include <stdio.h>
#define MAX_ROWS 5
#define MAX_COLS 8
#define MAX_LAYERS 2

void main(void)
{
int student_population[MAX_ROWS][MAX_COLS][MAX_LAYERS],
    row, col, layer;          /* Loop control variables */

/* Array Initialization and Loading Omitted */

for (row=0; row < MAX_ROWS; ++row)
  for (col=0; col < MAX_COLS; ++col)
    for(layer=0; layer<MAX_LAYERS; ++layer)
      printf("Element %d %d %d contains %d.\n",
             row, col, layer,
             student_population[row][col][layer]);
}
```

. . . .

This code segment will first print the number of freshman, male, math majors; then the freshman, female, math majors; then freshman, male, computer-science majors; freshman, female, computer-science majors; and so on.

Try It Modify the program in Example 7-8 so that the array is initialized by reading the values from standard input. Then calculate the total number of students, and print it on standard output.

7-5 STRUCTURES IN C

Earlier you learned that all the elements contained in an array must be of the same data type. A *structure* is a composite data type that allows you to store items of different data types or meanings in the same collection. For example, in the rain-sample program, suppose we wish to know the date and time each sample is taken in addition to its hydronium-ion concentration. The date is made up of three parts: the day, month, and year; the time is made up of two parts: the hour and minutes. Although each part has a different meaning, the parts must be kept together to have meaning. A structure allows you to keep these parts together.

To use a structure in a *C* program, you must define the structure as a new data type and declare variables of this new data type. To declare a structure to store the date, you use the *C* `struct` statement as follows:

```
struct date_type{
   int day;      /* Stores day value */
   int month;    /* Stores month value */
   int year;     /* Stores year value */
};
```

This `struct` statement appears near the top of your program and defines a new data type called `date_type` that has three fields, day, month, and year, each of which is an integer. A *structure field* is used to store the individual parts of the structure. The general form of a structure definition is

```
struct <struct name> {
  <field 1>
  <field 2>
       .
       .
       .
  <field n>
};
```

where

```
<field x> ::= <data type> <field name>;
```

This form tells you that a structure has a name and one or more fields.

Now that you have a new data type, you can declare variables of this type. The declaration of structure variables is preceded by the key word `struct` and looks similar to the declaration of scalar variables. To declare a variable of type `date_type` named `birthday`, you place the *C* statement

```
struct date_type birthday;
```

anywhere after the definition of `date_type`. The general form for declaring structure variables is

```
struct <structure type> <variable list>;
```

where

```
<variable list> ::= <variable name> [<variable list>];
```

Like any variable, `birthday` must be initialized with some meaningful data before it can be used in a program. The fields of a structure variable are accessed using the *dot operator* (.). To initialize the day field of `birthday` to 2, `month` to 7, and `year` to 1960, you would write three separate assignment statements as follows:

```
birthday.day = 2;
birthday.month = 7;
birthday.year = 1960;
```

You can also define a structure to represent time as hours and minutes with the `struct` statement.

```
struct time_type{
    int hour;    /* Stores hour part of the time */
    int minutes; /* Stores minutes past the hour */
}
```

Using the `date_type` and `time_type` structures, you can build a new structure to collect the hydronium-ion concentration of a rain sample along with the date and time the sample was taken. This is accomplished by placing structures within structures.

```
struct sample_type{
    struct date_type date;     /* Stores date sample was taken */
    struct time_type time;     /* Stores time sample was taken */
    float h_concentration;     /* Stores hydronium concentration */
    float ph_level;            /* Stores converted pH value */
};
```

Note that we have added a fourth field for the pH equivalent of the hydronium concentration. This extra field allows us to retain the raw data for possible use later.

It is perfectly acceptable for the fields of a structure to be structure variables or arrays. The following structure definition represents a student grade record and records the semester as an integer; a set of class grades is recorded in a `char` array:

```
struct student_grades{
    int semester;
    char grades[MAX_CLASSES];
}
```

It is possible for the elements of an array to be structures rather than scalar values. These capabilities add a great deal of programming power to C. Now the `rain_data` array can be redefined as

```
struct sample_type rain_data[MAX_RAINFALLS][MAX_SITES];
```

This declaration says that `rain_data` is a two-dimensional array of `MAX_RAINFALLS` rows by `MAX_SITES` columns, where each element is a `sample_type` structure. Example 7-9 contains part of a revision of the program in Example 7-7 that uses this declaration as well as a function to initialize the `rain_data` array. Note that we have left the declarations for the average arrays in place as well as their initializations. The remainder of this revision is an exercise for you to complete.

EXAMPLE 7-9   Revised Calculation of pH Statistics

```
#include <stdio.h>
#include <math.h>

#define MAX_SITES 8        /* Number of columns in array */
#define MAX_RAINFALLS 5    /* Number of rows in array */

struct date_type{
  int day;                 /* Stores day value */
  int month;               /* Stores month value */
  int year;                /* Stores year value */
};

struct time_type{
  int hour;                /* Stores hour part of the time */
  int minutes;             /* Stores minutes past the hour */
};

struct sample_type{
  struct date_type date;   /* Stores date sample was taken */
  struct time_type time;   /* Stores time sample was taken */
  float h_concentration;   /* Stores hydronium concentration */
  float ph_level;          /* Stores converted pH value */
};

/* Function Prototypes */
void init_array(
        struct sample_type array[MAX_RAINFALLS][MAX_SITES]);

void main(void)
{
  struct sample_type rain_data[MAX_RAINFALLS][MAX_SITES];
  float site_averages[MAX_SITES],
                        /* Stores separate site stats */
    rainfall_averages[MAX_RAINFALLS],
                        /* Stores separate rainfall stats */
    overall_average;    /* Average pH */

  int lcv, row, col;    /* Loop control variables */

  /* Initialize rain data array */
  init_array(rain_data);

  /* Initialize statistics arrays */
  for (lcv=0; lcv < MAX_RAINFALLS; ++lcv)
    rainfall_averages[lcv] = 0.0;
```

```
    for (lcv=0; lcv < MAX_SITES; ++lcv)
      site_averages[lcv] = 0.0;


    /* Remainder of revision left as an exercise */
}


/*------------------------------------------------------------------
Module: init_array()
Description: Initializes all fields in a sample_type array to 0;
Input Parms: A 2D array of sample_type structures.
Returns: none
Modifications
        Date            Description
        none            none
-----------------------------------------------------------------*/
void init_array(struct sample_type array[MAX_RAINFALLS][MAX_SITES])
{
    int row, col;           /* Loop control variables */

    for (row=0; row < MAX_RAINFALLS; ++row)
      for (col=0; col < MAX_SITES; ++col){
        array[row][col].date.day = 0;
        array[row][col].date.month = 0;
        array[row][col].date.year = 0;
        array[row][col].time.hour = 0;
        array[row][col].time.minutes = 0;
        array[row][col].h_concentration = 0.0;
        array[row][col].ph_level = 0.0;
      }
}
```

. . . .

Carefully examine the definition of init_array(). Notice that the parameter array is declared as a two-dimensional array of structures. In the body of this function access is made to elements within array using the familiar subscript notation. Once a particular element has been accessed in this manner, the fields of that structure are accessed using the dot operator. Whenever the accessed field is itself a structure, its fields are accessed using a second dot operator, as in the statement

```
array[row][col].date.month = 0;
```

Because the dot is an operator, it can be combined with other operators in this way.

There are many powerful ways to use structures in a program, and this section provides you with only an introduction. In the following application you will learn how structures can be used to extend the mathematical power of *C* to include complex arithmetic.

What If   **?**   Write a simple program using date_type to calculate a person's age in years, months, and days by subtracting his or her birthdate from the current date.

Application 1   FREQUENCY DISTRIBUTION GRAPHING

Industrial Engineering

Quality-control engineers monitor the quality of an automated production line by tracking the number of defective parts coming off the line within a particular period. If the frequency of defective parts rises dramatically for a given period, the engineer is alerted that a problem exists and can take action to fix the problem. Such frequencies can be depicted using a bar graph such as the one shown in Figure 7-5. In this graph, the horizontal axis represents the number of defects detected and the vertical axis represents data collection periods.

You can automate this frequency reporting by writing a program that collects defect data every half-hour and stores the frequencies as array elements. This data can then be used to print a frequency distribution graph on the computer's screen.

 ## 1. Define the Problem

The problem is how to write a program that will read the number of defective parts for each half-hour in a single day and store them as elements in a one-dimensional array. This array is to be used to display a frequency distribution graph on the computer's display monitor.

 ## 2. Gather Information

The input data for this program is a frequency count of defective parts for each half-hour beginning with 12:00 midnight on one day and running to 12:00 midnight of the following day, consecutively. Hence there

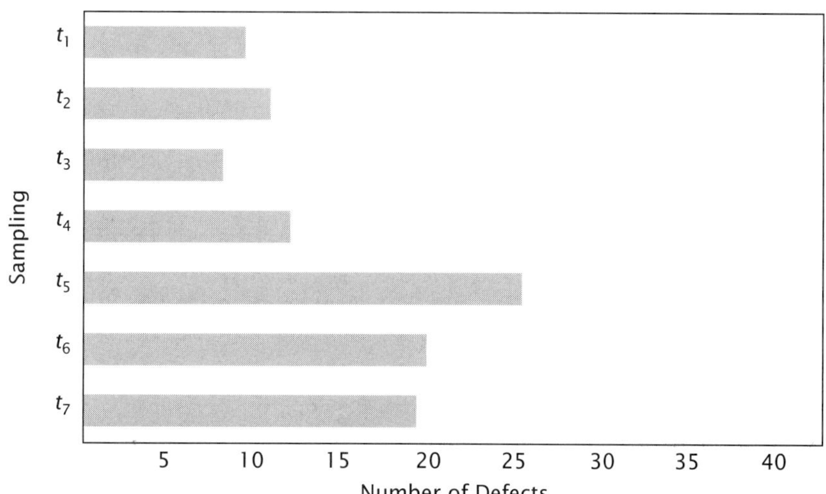

Figure 7-5 Frequency Distribution Graph

will be 48 frequency counts, each of which will be a single integer value. These inputs will be entered on standard input by the inspectors on the production line and should not exceed 50, because 50 parts are produced every half-hour. Additionally, negative inputs should be considered invalid.

The output data for this program is a frequency distribution graph. The graph will resemble the graph shown in Figure 7-5 in layout, however, the graph will be generated using text characters.

3. Generate and Evaluate Potential Solutions

The problem can be decomposed into two general parts: reading frequency data and displaying the frequency graph. Although there are 48 data collections in a given day, it is possible that frequency data is available for only some of these periods. Therefore, you must design your program so that it can partially fill the array with frequency data. You can do this by prompting for the number of periods for which data exists. This value is read from standard input prior to reading the frequency data. So, reading frequency data requires the following steps:

```
Read and check number of periods for which data is available
Read and check each frequency data value
```

The structure chart shown in Figure 7-6 reflects the design of a program to solve this problem.

You will handle reading and checking input similarly to the approaches in previous examples. The algorithm for reading the number of data values is as follows:

```
Do
    Prompt user for number of data values
    Read number
While number < 0 or > 48
```

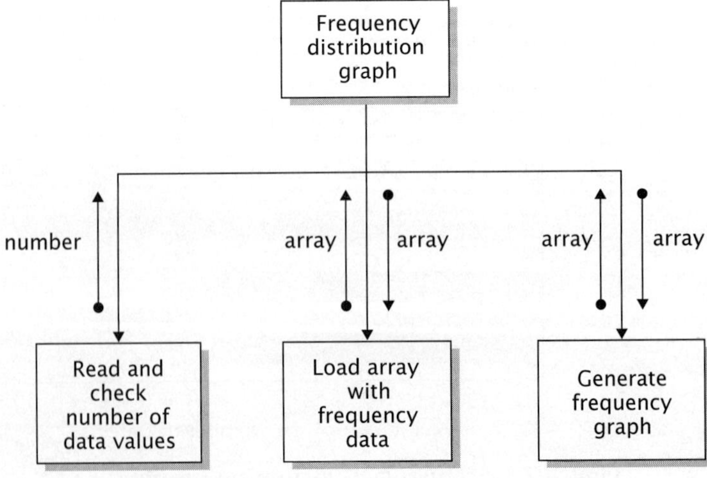

Figure 7-6 Structure Chart for Distribution Plotting

The algorithm for loading the array will use number from the previous algorithm as its boundary value. Each value entered must be checked for validity.

```
For each frequency data item
   Do
      Prompt user for frequency data
      Read value into frequency array
   While value < 0 or > 50
```

The printing of the frequency distribution graph requires some additional thought. This algorithm must print the vertical axis, each frequency bar, and the horizontal axis. Because the bars are printed horizontally, you can simply traverse the array and print a bar to represent each value. You can print the bars as a sequence of asterisks or some other visually appealing character. Using this approach, the graph might appear as follows:

```
     --------------------------------------------------
 1 | *************************
 2 | **************************
 3 | ************************
 4 | **************
 5 | ***********************
 6 | ********************
 7 | **********************
 8 | ****
 9 | ******************
10 | *******************
     --- 5---10---15---20---25---30---35---40---45---50
```

This output is much like printing a two-dimensional array. You need a for loop for each row and one for each column. An algorithm to solve this problem is as follows:

```
Print the graph heading
For each frequency array element
   Print part of the vertical axis
   For each value from 0 to the frequency value
     Print an asterisk
   Move cursor to next line
Print the horizontal axis
```

You will develop each of the previous algorithms into a *C* function definition in the following step.

◢ 4. Refine and Implement a Solution

The structure chart and algorithms developed in the previous section are implemented as the program in Example 7-10.

EXAMPLE 7-10   Plotting Frequency Distributions on a Bar Graph

```c
/*-------------------------------------------------------------------
   Description: This program prints the frequency of defective
      parts in bar chart format.
   Programmer: Ken Collier
   Date of Last Revision: 1-1-1995
   Modifications
         Date              Description
         none              none
-------------------------------------------------------------------*/
#include <stdio.h>

#define MAX_SIZE 48
#define MAX_PARTS 50

/* Function Prototypes */
int get_number_of_values(void);
void load_array(int data[MAX_SIZE], int upper_bound);
void generate_graph(int data[MAX_SIZE], int upper_bound);

void main(void)
{
  int data_cnt,                  /* Number of data values */
     frequencies[MAX_SIZE];      /* Defect frequency values */

  data_cnt = get_number_of_values();
  load_array(frequencies, data_cnt);
  generate_graph(frequencies, data_cnt);
}


/*-------------------------------------------------------------------
Module: get_number_of_values()
Description: Reads and checks the number of data items to be
input. Value must be in the range 0 <= number <= 48.
Input Parms: none
Returns: The number of data items to be entered.
Modifications
         Date              Description
         none              none
-------------------------------------------------------------------*/
int get_number_of_values(void)
{
  int number;                    /* Number of data values */

  /* Read input and check for errors */
  do{
    printf("Enter number of frequency values (0-48): ");
    scanf("%d", &number);
  }while (number < 0 || number > 48); /* While input error */

  return(number);
}
```

```
/*----------------------------------------------------------------
Module: load_array()
Description: Loads the frequency array with the counts of
  defective parts for each time period. Values must be in the
  range 0 <= number <= 50.
Input Parms: frequency array, number of data items.
Returns: loaded frequency array.
Modifications
        Date                Description
        none                none
----------------------------------------------------------------*/
void load_array(int data[MAX_SIZE], int upper_bound)
{
  int lcv;                        /* Loop control variable */

  /* Load values into array elements */
  for(lcv=0; lcv < upper_bound; ++lcv)
    /* Read input and check for errors */
    do{
      printf("Enter data for time period %d (0-50): ", lcv+1);
      scanf("%d", &data[lcv]);
    }while (data[lcv] < 0 || data[lcv] > 50);
}


/*----------------------------------------------------------------
Module: generate_graph()
Description: Displays bar graph with horizontal bars showing the
  defective frequencies for every half-hour beginning with 12:00
  midnight.
Input Parms: frequency array, number of data items.
Returns: none
Modifications
        Date                Description
        none                none
----------------------------------------------------------------*/
void generate_graph(int data[MAX_SIZE], int upper_bound)
{
  int bar,                /* Outer loop control variable */
      length,             /* Inner loop control variable */
      lcv;                /* General purpose lcv */

  /* Print bounding box on graph */
  printf("\n\n     ");
  for(lcv=0; lcv < 50; ++lcv)
    putchar('-');
  putchar('\n');
```

```c
/* Print graph body */
for(bar=0; bar < upper_bound; ++bar){
  /* Print vertical axis */
  printf("%2d | ", bar + 1);
  /* Print frequency bar */
  for (length = 0; length < data[bar]; ++length)
    putchar('*');
  putchar('\n');
}

/* Print the horizontal axis */
printf("     ");/* Proper spacing */
for(lcv=5; lcv < 55; lcv += 5)
  printf("---%2d", lcv);
putchar('\n');
}
```

. . . .

5. Verify and Test the Solution

To properly test this program, you should enter a variety of values for
first input, including 0, 48, values below 0, values above 48, and valid
values. Selecting values along the boundaries is known as boundary value
testing. You should do the same for the actual frequency values. Given
the input values −3 (error) 55 (error) 10−9 (error) 15 16 13 14 20 9 96
(error) 19 14 19 20, the output of this program is

```
      --------------------------------------------------
 1 | ***************
 2 | *****************
 3 | *************
 4 | **************
 5 | ********************
 6 | *********
 7 | ********************
 8 | **************
 9 | *********************
10 | *********************
      --- 5---10---15---20---25---30---35---40---45---50
```

What If   Modify the distribution plotting program so that it accepts readings from
15-minute periods. Because the period length is halved, the maximum
possible defective products per period will also be cut in half. This reduction will cause most of the bars to be very short. To gain greater resolu-
tion, modify the program so that each increment is represented by two
stars instead of one.

SUMMARY

This chapter introduced you to one-, two- and three-dimensional arrays, as well as structures. Arrays and structures offer a cohesive means of storing composite collections of data. Arrays are used for grouping items of the same type and meaning, whereas structures allow you to group related items with different types and meanings. You learned how to declare and initialize arrays and manipulate the elements in an array. You also learned how to define a structure, declare structure variables, and manipulate the fields in a structure. Finally, you learned that these composite data types can be combined to allow you to create arrays of structures, structures containing arrays, and structures containing other structures. Because most programs today manage large amounts of data, these tools provide you with important programming power.

Key Words

array	row-major order
array element	subscript notation
array index	storage cells
array name	structure
array size	structure field
array subscript	three-dimensional array
dot operator	two-dimensional array
one-dimensional array	

Exercises

1. Write a *C* program that declares a ten-element array named `voltages`. Your program should read a set of values into the array, call a function called `average()` to calculate the average voltage, and print all voltages followed by the average.

2. Modify your solution to problem 1 so that it also calls a function called `variance()` that calculates the variance of the voltages. Variance is calculated by subtracting the average from each original voltage value, squaring these values, adding them together, and dividing by the number of numbers. Print the variance below the average.

3. In Example 7-2 you saw a program that used two separate loops for initializing the odd cells of an array to 5 and the even cells to 10. Another alternative is to visit each cell in a single loop. If the index is even, initialize the cell to 10, otherwise, initialize the cell to 5. This initialization requires an `if-else` statement nested within a `for` loop. Modify the program in Example 7-2 that implements this approach.

4. The maximum value in a collection of values can be found by applying the algorithm

```
set max to the first array element
for each element in the array
   if the current element is greater than max then
      set max to the current element
```

Write a *C* program that implements this algorithm for a collection of ten integers.

5. Write a *C* program that processes parts orders for the shipping department at ACME Engineering. The program declares 4 one-dimensional arrays to hold 10 values. The first array stores part numbers as integers; the second stores prices as real numbers; the third stores quantity as integers; and the fourth stores amounts as real numbers. These arrays are processed in parallel. For example, `part_num[0]` contains a part number for a part whose cost is stored in `price[0]`. The quantity of this part ordered by the customer is stored in `quantity[0]` and `total[0] = price[0] * quantity[0]`. Design your program to read part number, price, and quantity for each part from standard input and to display the total order in table form, like this:

Part	Price	Quantity	Amount
352	0.75	500	375.00
355	1.30	300	390.00
456	4.25	1000	4250.00
.			
.			
.			

6. Rework your solution to problem 5 to use an array of structures instead of four separate arrays. The structure definition should have four fields for part number, price, quantity, and amount.

7. Modify the acid-rain statistics program in Example 7-7 so that the actions of loading the array and calculating the statistics are performed in functions. Pass the `rain_data`, `site_averages`, and `rainfall_averages` arrays as a parameter to those functions.

8. Modify the acid-rain statistics program in Example 7-7 to use an array of `sample_type` structures. Design your program to print the pH level of each sample as it is entered; calculate the average hydronium-ion concentration.

9. One method of smoothing variations in sequential data is the use of a moving average filter. This filter simply takes the last three to five data points, averages them, and then reports the average. The program then moves ahead one data point and repeats the process. If a data acquisition system is supplying the following voltage readings, taken every 10 seconds,

$$5.5 \quad 3.4 \quad 4.3 \quad 5.2 \quad 5.1 \quad 4.6 \quad 3.4 \quad 4.5 \quad 4.7 \quad 5.6$$
$$5.8 \quad 6.0 \quad 5.5 \quad 5.0 \quad 4.6 \quad 4.3 \quad 5.0 \quad 5.4 \quad 5.7 \quad 5.9$$

write a program to calculate and print the moving average using an averaging period of three data points. What does the series of averages look like? Can you think of an application where you might want to smooth the data?

10. Repeat problem 9 using an averaging period of five data points. What happens to the average values? How can you handle the last four readings?

11. Many manufacturing industries have adopted statistical process control (SPC) as a useful technique to implement quality control. Write a program to evaluate input data and print "The process has exceeded the limit" for values above a limit of 10 and print "The process is below the limit" for values below 5, where the data is

$$6 \quad 7 \quad 6 \quad 5 \quad 4 \quad 8 \quad 9 \quad 8 \quad 7 \quad 10 \quad 11 \quad 9 \quad 8 \quad 6 \quad 5$$
$$4 \quad 4 \quad 6 \quad 7 \quad 8 \quad 9 \quad 8 \quad 7 \quad 8 \quad 6 \quad 7 \quad 8 \quad 8 \quad 9 \quad 7$$

How could you use this program during the production process to provide operator feedback?

12. Engineers often rely on tables for calculations, and even today tables have their place for performing rapid calculations to obtain an approximate answer. Write a program utilizing a two-dimensional array to calculate a table that can be used to quickly look up pH values given the molar concentration of the hydrogen ion. Structure the table where pH is given to one decimal place and ranges from 2 to 10 (2.0, 2.1, 2.2, . . . , 9.8, 9.9, 10.0).

13. Frequently, engineers are not concerned with the actual value of an event but with the number of times that event has occurred. One area where this happens is in determining the frequency of rainfall events and estimating the probability of occurrence of a storm of a given size. These calculations are vital in sizing reservoirs, storm water sewers, channels, and other water-control structures. The first step in this process is to count the number of times a storm of a particular size has occurred for a given set of data. Using the rainfall data given in the table, write a program that counts the number of storms that are greater than or equal to 60 mm/hr.

Annual Average Rainfall Intensities

Year	Intensity (mm/hr)
1984	95.3
1985	82.9
1986	53.4
1987	47.8
1988	43.2
1989	59.8
1990	44.6
1991	59.9
1992	71.5
1993	68.5

14. Expand the program written for problem 13 to count the number of values from 40 to 50, 50 to 60, 60 to 70, 70 to 80, and 80 to 90. If the data was for a period of 100 years, how would this change your program?

15. Digital images are made up of a two-dimensional array with cell values that represent the color, hue, and intensity. Write a program in which an

array is used to represent your school logo in black and white, where the value of 1 in the cell represents black and 0 represents white. How does the dimension of the array affect the quality of the picture?

16. The quality of a body of water is determined in part by the amount of dissolved oxygen available for aquatic life. In streams and rivers, discharge of an organic pollutant will result in a decrease in the amount of dissolved oxygen downstream. This dissolved oxygen reduction can be described by the equation

$$D = \frac{k \times Lo}{kr - k} \times [\exp(-k \times t) - \exp(-kr \times t)] - [Da \times \exp(-kr \times t)]$$

where

D = the oxygen deficit in (mg/L) at time t (day)
k = the oxygen depletion constant (1/day)
Lo = the amount of pollutant (mg/L)
kr = the re-aeration constant (1/day)
Da = the initial dissolved oxygen deficit (mg/L)

Write a program using a one-dimensional array with 100 cells. Considering each cell as 0.1 day long and starting with the first cell as time = 0 days, calculate the value of D for each cell (that is, as time moves forward 0.1 day at a time, calculate the oxygen deficit). Let

k = 0.35 1/day
Lo = 200 mg/L
kr = 0.4 1/day
Da = 3.5 mg/L

17. Use the program written for problem 16 to determine and print where the maximum oxygen deficit occurs.

18. Many engineering instruments utilize digital signals in reporting their measurements. In developing a digital signal, an analog signal is often converted to digital form by an A/D conversion. Using a one-dimensional array, write a program to digitize a 60Hz analog sine wave. The sine wave can be described by the equation

$$V(t) = \sin(2\pi 60t)$$

where $V(t)$ is the voltage at time t, given in seconds. *Hint:* Consider each cell in the array as one little segment of time, calculate the voltage at that time, and increment to the next cell.

19. Define a `struct` type called `water_sample_type` that includes
    1. Technician—Stores an integer ID number
    2. Sampling location—Stores an integer ID number
    3. Sampling date—Use the `date_type` defined previously
    4. Sampling time—Use the `time_type` defined previously
    5. Water temperature
    6. pH
    7. Dissolved oxygen

20. Using your definition in problem 19, write a program that declares and initializes a one-dimensional array named `lake_samples` in which each element is a `water_sample_type`. Read water sample data from standard input, and calculate and print minimums, maximums, and averages for each of the sample values.

Appendix A
ASCII Table

Table A-1 contains the characters associated with the American Standard Code for Information Interchange (ASCII) decimal codes. The first 30 characters are either nonprinting control characters or semiprinting characters (such as, space, tab, and newline). Some important abbreviations in this table are HT, LF, FF, CR, ESC, and DEL. These represent the characters horizontal tab, line feed, form feed, carriage return, escape, and delete, respectively.

Table A-1   American Standard Code for Information Interchange (ASCII)

Digits Left / Right	0	1	2	3	4	5	6	7	8	9
0	Null	SOH	STX	ETX	EOT	ENQ	ACK	Bell	BS	HT
1	LF	VT	FF	CR	SO	SI	DLE	DC1	DC2	DC3
2	DC4	NAK	SYN	ETB	CAN	EM	SUB	ESC	FS	GS
3	RS	US	Space	!	"	#	$	%	&	'
4	(	)	*	+	,	−	.	/	0	1
5	2	3	4	5	6	7	8	9	:	;
6	<	=	>	?	@	A	B	C	D	E
7	F	G	H	I	J	K	L	M	N	O
8	P	Q	R	S	T	U	V	W	X	Y
9	Z	[	\	]	∧	−	`	a	b	c
10	d	e	f	g	h	i	j	k	l	m
11	n	o	p	q	r	s	t	u	v	w
12	x	y	z	{	|	}	~	DEL		

Appendix B
File Input and Output

Throughout this module your programs have been reading from standard input (keyboard) and writing to standard output (display monitor). Often the data input to a program resides in an electronic file. Your program can read these input files in much the same way you have been reading input from the keyboard. Furthermore, you may want the output from your program to be stored in an electronic output file for future reference. You can design your programs to write to such a file instead of writing to the display monitor. Most large commercial programs make heavy use of files rather than requiring a human operator to enter all data manually. This appendix will extend your programming abilities by giving you an introduction to file input and output.

Conceptually, file I/O is very similar to standard I/O. When you write to a file you use a function called `fprintf()`, and when you read from a file you use a function called `fscanf()`. These functions are very similar to their standard I/O counterparts `printf()` and `scanf()`; however, before you read from or write to a file, you must explicitly open the file reading, writing, or both. After you are finished with the file you must close it. We will discuss each of these concepts and provide examples.

OPENING A FILE

In *C* a file may be any of several different entities. Typically a file refers to an electronic file on disk. A file it may also refer to a printer or any other device. In fact, the keyboard and the display monitor can be thought of as files. In this discussion we will view files in their more traditional interpretation, as electronic data files on a disk drive.

You first must associate your program with a file by opening the file. Once the file is opened you can send information to the file or get information from the file. To do this your program must have a means of referring to the file. This reference is accomplished using a file pointer. A *file pointer* is a pointer variable of type `FILE` that is defined in `stdio.h`. To declare a

Table B-1   Text File Modes

Mode	Meaning
r	Open a text file for reading
w	Create a text file for writing
a	Open a text file for appending
r+	Open a text file for reading and writing
w+	Create a text file for reading and writing
a+	Open or create a text file appending, reading and writing

variable as a file pointer you use the statement

```
FILE *file_ptr;
```

where the name `file_ptr` can be any name you wish to use to refer to the file.

Once you have declared a file-pointer variable, you are ready to open a file. You open a file using the `stdio` function `fopen()`. This function has the prototype

```
FILE *fopen(char *filename, char *mode);
```

This prototype tells you that you must specify the name of the file and a mode under which the file will be manipulated. This function returns a file pointer, which should be stored in the file-pointer variable you declared previously.

When you open a file in a *C* program, you must define the mode of the file or the types of operations that will be performed on the file. If a file already exists, it can be read from, written over, or appended to, whereas a file that doesn't exist yet must be created and written to. After a file has been created, it can be read from and appended to. Table B-1 shows the possible modes under which a text file may be opened.

Suppose you have an input data file on your hard-disk drive named `raw-data.txt` that you wish to open for reading data into your program. The code segment in Example B-1 shows how to open this file.

EXAMPLE B-1

Opening a Text File for Reading

```
#include <stdio.h>

void main(void)
{
  FILE *infile;

  infile = fopen("raw-data.txt", "r");
        .
        .
        .
```

Following the call to fopen(), the variable infile points to the file raw-data.txt and can be used to access the contents of the file.

If the file is unable to be opened successfully, fopen() returns a null pointer. A null pointer's value is equivalent to the symbolic constant NULL, which is defined in stdlib.h. The code segment in Example B-2 shows how you can use this constant to check for file-opening errors in your program.

Checking for File-Opening Errors

```c
#include <stdio.h>

void main(void)
{
  FILE *infile;

  if ((infile = fopen("raw-data.txt", "r")) == NULL){
    printf("File error.\n");
    exit(1);
  }
    .
    .
    .
```

Notice that the file is opened directly inside the if statement header. This is common practice but may make the code less readable to the novice programmer.

READING FROM AND WRITING TO A FILE

Once a file is opened properly it can be manipulated according to the mode under which it was opened. The two most common functions for reading from and writing to a file are fscanf() and fprintf(), respectively. These functions reside in the stdio library and have the following prototypes:

```c
int fprintf(FILE *file_ptr, char *control_string, ...);
```

and

```c
int fscanf(FILE *file_ptr, char *control_string, ...);
```

These functions behave exactly like their standard I/O counterparts except that they operate on files and require a file-pointer variable as their first argument.

Suppose the file raw-data.txt contains a series of integer data values to be read as input into your program. To read a single value from this file into a variable called data you would use the statement

```c
fscanf(infile, "%d", &data);
```

Generally you will want to read all the data from a file. Because you may not know how many data values reside in a particular file, *C* provides a function called feof() that helps you check for the end of a file (EOF). This function has the prototype

```
int feof(FILE *file_ptr);
```

and returns true (nonzero value) if the end of file is reached, false (0) otherwise. You can use feof() to control a while loop that reads the next data item on each iteration. Consider the program segment in Example B-3 that reads integers from raw-data.txt into an integer array. This segment assumes there are no more than 1000 data values in the file.

EXAMPLE B-3   ## Reading Data until EOF

```
#include <stdio.h>

void main(void)
{
  FILE *infile;
  int element = 0, data[1000];

  if ((infile = fopen("raw-data.txt", "r")) == NULL){
    printf("File error.\n");
    exit(1);
  }

  while ((feof(infile)==0) && (element < 1000))
    fscanf(infile, "%d", &data[element++]);
       .
       .
       .
```

. . . .

In addition to the input file raw-data.txt, you might wish to write to an output file named final-data.txt. As with the first file, this file must be opened with its own file pointer. You can then use fprintf() to write data to the file. The incomplete program in Example B-4 shows how you can read from raw-data.txt and write to final-data.txt. Note that because the output file does not exist prior to the program's execution, the file mode "w" *creates* a file for writing.

EXAMPLE B-4   ## Writing to a File

```
#include <stdio.h>
#define MAX_SIZE 1000

void main(void)
{
  FILE *infile, *outfile;
  int lcv, element = 0, data[MAX_SIZE];
```

```c
   /* Open files */
   if ((infile = fopen("raw-data.txt", "r")) == NULL){
     printf("File error.\n");
     exit(1);
   }
   if ((outfile = fopen("final-data.txt", "w")) == NULL){
     printf("File error.\n");
     exit(1);
   }

   /* Read all data from raw-data.txt */
   while ((feof(infile)==0) && (element < MAX_SIZE))
     fscanf(infile, "%d", &data[element++]);

   /* Process values in data array */

   /* Write data to final-data.txt */
   for(lcv = 0; lcv < element; ++lcv)
     fprintf(outfile, "%d\n", data[element]);
            .
            .
            .
}
```

. . . .

Although this program does not perform any processing on the data, typically you will read the input data into a composite data structure such as an array, process the elements of the array, and write the resulting data to an output file.

CLOSING A FILE

After the processing of a file is complete, it must be closed. Among other operating system activities, closing a file places an EOF marker at the end of the file. To close a file you use the *C* function `fclose()`, which has the prototype

```c
int fclose(FILE *file_ptr);
```

A return value of 0 indicates that the file was closed successfully. Any other return value indicates a file-closing error.

Using this function, the program in Example B-4 must end with the statements

```c
fclose(infile);
fclose(outfile);
```

pH FACTORS REVISITED

Recall Example 7-7 in which we calculated pH averages for several rainfall samples. This program was designed to read input data from the keyboard and display output to the display monitor. Example B-5 shows a modification of the same program in which rain-sample data is read from a file called `raw-data.txt` and is written to an output file called `final-data.txt`. The boldface statements in the program are the new or modified lines of code. Note that there were very few modifications to the program.

EXAMPLE B-5

Calculating pH Statistics using File I/O

```c
/*------------------------------------------------------------------
   Description: This program calculates and prints:
      - The overall average acidity of 8 rainfall samples
        taken over 5 rainfalls.
      - The average acidity per rainfall of all sites.
      - The average acidity per site over 5 rainfalls.

      Acidity is measured on a pH scale. pH factors
      are calculated using the relationship

          pH = -log10(moles of hydronium ions/liter)

   Programmer: Ken Collier
   Date of Last Revision: 1-1-1995
   Modifications
         Date              Description
         none              none
   --------------------------------------------------------------*/

#include <stdio.h>
#include <math.h>

/* Symbolic Constants */

#define MAX_SITES 8/* Number of columns in array */
#define MAX_RAINFALLS 5/* Number of rows in array */

/* Function Prototypes */

void init_array(float data[MAX_RAINFALLS][MAX_SITES]);
void read_and_convert_rain_data(
                float data[MAX_RAINFALLS][MAX_SITES]);
float calc_overall_average(float data[MAX_RAINFALLS][MAX_SITES]);
void calc_average_per_rainfall(
                float data[MAX_RAINFALLS][MAX_SITES],
                float avgs[MAX_RAINFALLS]);
void calc_average_per_site(float data[MAX_RAINFALLS][MAX_SITES],
                float avgs[MAX_SITES]);
void output_site_averages(FILE *fp, float data[MAX_SITES]);
void output_rainfall_averages(FILE *fp, float data[MAX_RAINFALLS]);
```

```c
void main(void)
{
  float       rain_data[MAX_RAINFALLS][MAX_SITES],
              site_averages[MAX_SITES],
                  /* Stores separate site stats */
              rainfall_averages[MAX_RAINFALLS],
                  /* Stores separate rainfall stats */
              overall_average;/* Average acidity */
  int lcv;                    /* Loop control variable */
  FILE *outfile;              /* File pointer to output file */

  /* Initialize all arrays */
  init_array(rain_data);

  for (lcv=0; lcv < MAX_RAINFALLS; ++lcv)
    rainfall_averages[lcv] = 0.0;

  for (lcv=0; lcv < MAX_SITES; ++lcv)
    site_averages[lcv] = 0.0;

  read_and_convert_rain_data(rain_data);

  /* Calculate statistics */
  overall_average = calc_overall_average(rain_data);
  calc_average_per_rainfall(rain_data, rainfall_averages);
  calc_average_per_site(rain_data, site_averages);

  /* Output overall average, site averages, & rainfall averages */
  /* First open output file */
  if ((outfile = fopen("final-data.txt", "w")) == NULL){
    printf("File error.\n");
    exit(1);
  }

  /* Now write output to file */
  fprintf(outfile, "\nThe overall average pH is %.2fpH.\n\n",
          overall_average);
  output_site_averages(outfile, site_averages);
  output_rainfall_averages(outfile, rainfall_averages);
  /* Close output file */
  fclose(outfile);
}


/*----------------------------------------------------------------
Module: init_array()
Description: Initializes a 2D array.
Input Parms: A 2D array of MAX_RAINFALLS x MAX_SITES
Returns: None
Modifications
        Date            Description
        none            none
----------------------------------------------------------------*/
```

```c
void init_array(float data[MAX_RAINFALLS][MAX_SITES])
{
   int row, col;/* Loop control variables */

   /* Initialize rain data array */
   for (row=0; row < MAX_RAINFALLS; ++row)
     for (col=0; col < MAX_SITES; ++col)
       data[row][col] = 0.0;
}


/*-------------------------------------------------------------------
Module: read_and_convert_rain_data()
Description: Loads a 2D array with rain data from an input file
  and converts it to pH factor.
Input Parms: A 2D array of MAX_RAINFALLS x MAX_SITES
Returns: None
Modifications
        Date           Description
        none           none
-------------------------------------------------------------------*/
void read_and_convert_rain_data(
                      float data[MAX_RAINFALLS][MAX_SITES])
{
  int row, col;/* Loop control variables */
  char site;   /* For printing site labels */
  FILE *infile;/* File pointer to input file */

  /* Open input data file */
  if ((infile = fopen("raw-data.txt", "r")) == NULL){
    printf("File error.\n");
    exit(1);
  }

  /* Read rainwater data from standard in & convert to pH scale */
  for (row=0; row < MAX_RAINFALLS; ++row){
    site = 'A';
    for (col=0; col < MAX_SITES; ++col){
      fscanf(infile, "%f", &data[row][col]);
      data[row][col] = -log10(data[row][col]);
    }
  }

  /* Now close input file */
  fclose(infile);
}


/*-------------------------------------------------------------------
Module: calc_overall_average()
Description: Averages the pH factors in a 2D array.
Input Parms: A 2D array of MAX_RAINFALLS x MAX_SITES
Returns: None
Modifications
        Date           Description
        none           none
-------------------------------------------------------------------*/
```

```c
float calc_overall_average(float data[MAX_RAINFALLS][MAX_SITES])
{
  int row, col;   /* Loop control variables */
  float total=0.0;/* Running total */

  /* Calculate overall average acidity */
  for (row=0; row < MAX_RAINFALLS; ++row)
    for (col=0; col < MAX_SITES; ++col)
      total += data[row][col];

  return((float)total / ((float)MAX_RAINFALLS * (float)MAX_SITES));
}


/*------------------------------------------------------------------
Module: calc_average_per_rainfall()
Description: Averages each row in a 2D array.
Input Parms: A 2D array of MAX_RAINFALLS x MAX_SITES
Returns: None
Modifications
        Date              Description
        none              none
------------------------------------------------------------------*/
void calc_average_per_rainfall(
                    float data[MAX_RAINFALLS][MAX_SITES],
                    float avgs[MAX_RAINFALLS])
{
  int row, col;   /* Loop control variables */
  float total=0.0;/* Running total */

  /* Calculate the average of each row */
  for (row=0; row < MAX_RAINFALLS; ++row){
    total = 0.0;
    for (col=0; col < MAX_SITES; ++col)
      total += data[row][col];
    avgs[row] = total / MAX_SITES;
  }
}


/*------------------------------------------------------------------
Module: calc_average_per_site()
Description: Averages each column in a 2D array.
Input Parms: A 2D array of MAX_RAINFALLS x MAX_SITES
Returns: None
Modifications
        Date              Description
        none              none
------------------------------------------------------------------*/
void calc_average_per_site(float data[MAX_RAINFALLS][MAX_SITES],
                  float avgs[MAX_SITES])
{
  int row, col;   /* Loop control variables */
  float total=0.0;/* Running total */
```

```c
    /* Calculate the average of each column */
    for (col=0; col < MAX_SITES; ++col){
      total = 0.0;
      for (row=0; row < MAX_RAINFALLS; ++row)
        total += data[row][col];
      avgs[col] = total / MAX_RAINFALLS;
    }
}


/*------------------------------------------------------------------
Module: output_site_averages()
Description: Prints each element in 1D array to a file.
Input Parms: A 1D array of MAX_SITES elements.
Returns: None
Modifications
        Date            Description
        none            none
-----------------------------------------------------------------*/
void output_site_averages(FILE *fp, float data[MAX_SITES])
{
  int lcv;  /* Loop Control Variable */
  char site;/* For printing site label */

  /* Output site averages */
  site = 'A';
  for (lcv=0; lcv < MAX_SITES; ++lcv)
    fprintf(fp, "The average for site %c is %.2fpH.\n",
        site++, data[lcv]);
  printf("\n");
}


/*------------------------------------------------------------------
Module: output_rainfall_averages()
Description: Prints each element in 1D array to a file.
Input Parms: A 1D array of MAX_RAINFALLS elements.
Returns: None
Modifications
        Date            Description
        none            none
-----------------------------------------------------------------*/
void output_rainfall_averages(FILE *fp, float data[MAX_RAINFALLS])
{
  int lcv;  /* Loop Control Variable */
  char site;/* For printing site label */

  /* Output rainfall averages */
  for (lcv=0; lcv < MAX_RAINFALLS; ++lcv)
    fprintf(fp, "The average for rainfall %d is %.2fpH.\n",
        lcv+1, data[lcv]);
}

. . . . . . . . . . . . . . . . . . . . . . . . . . . . . . . . . . . .
```

The input data file containing the values shown in Figure 7-2 would look like Example B-6.

EXAMPLE B-6   ## Contents of `raw-data.txt` Input File

```
.00025
.00004
.0003
.00174
.0008
.00001
.0009
.00045
.0003
.00003
.00022
.002
.00055
.0001
.0007
.0004
.0002
.000042
.00035
.0015
.00058
.00008
.00085
.00042
.00028
.000035
.00032
.00221
.00083
.00005
.0013
.00048
.00022
.000045
.00039
.0018
.00055
.00009
.00088
.0005
```

. . . .

Given this input file, when executed the program will generate the output file shown in Example B-7.

EXAMPLE B-7   ## Contents of `final-data.txt` after Execution

```
The overall average pH is 3.52pH.

The average for site A is 3.61pH.
The average for site B is 4.42pH.
The average for site C is 3.51pH.
The average for site D is 2.74pH.
The average for site E is 3.19pH.
The average for site F is 4.29pH.
The average for site G is 3.04pH.
The average for site H is 3.35pH.
The average for rainfall 1 is 3.60pH.
The average for rainfall 2 is 3.53pH.
The average for rainfall 3 is 3.52pH.
The average for rainfall 4 is 3.47pH.
The average for rainfall 5 is 3.48pH.
```

. . . .

This output data is consistent with the output from the program in Example 7-7.

SUMMARY

This appendix has introduced only a few of the file I/O features supported by ANSI *C*. As you write more complex programs you will probably make greater use of input and output data files. This approach allows you to create one program whose output data can act as the input data to another program, greatly increasing your programming power.

Index